The Last
Billion Years

The Last Billion Years

A Geologic History of Tennessee

Don W. Byerly

The University
of Tennessee Press

Knoxville

The paper in this book meets the requirements of American National Standards
Institute / National Information Standards Organization specification Z39.48-1992
(Permanence of Paper). It contains 30 percent post-consumer waste and is certified
by the Forest Stewardship Council.

Library of Congress Cataloging-in-Publication Data

Byerly, Don W.
The last billion years: a geologic history of Tennessee / Don W. Byerly. — First edition.
    pages cm
Includes bibliographical references and index.
ISBN 978-1-57233-974-3 (pbk.) — ISBN 1-57233-974-8 (pbk.)

1. Geology—Tennessee.
2. Sequence stratigraphy.
3. Geology, Stratigraphic.
I. Title.

QE165.B94 2013
557.68—dc23
2013001461

for Sue

Contents

Illustrations

Preface

The intent of this book is to present the tools and background useful for interpreting geologic history. And then, to present the geologic history of Tennessee in a fashion that can be easily understood by the general public as well as middle school, high school, and college students. The book also can serve as a handy reference for professional geologists who want a glimpse of Tennessee geology. The title, *The Last Billion Years: A Geologic History of Tennessee,* was chosen because the oldest rocks in Tennessee have been dated as being just a bit older than one billion years, and most of the geologic history preserved in Tennessee's rocks dates from about 800 million years ago to the present.

Although vast amounts of data relevant to the geologic history of Tennessee are available in technical and scientific journals, this book, while tapping many of those resources, tries to present the history in a more generalized manner. I apologize to those readers who detect that I have certain biases in the telling, but I confess a personal perspective that reflects my professional experiences. This book emphasizes the physical geologic history rather than serving as a compendium on the fossil record of Tennessee. It is anticipated that a companion book will be published in the future that will address the biological evolution through the annals of Tennessee's geologic history.

All sciences use terms that can dampen the enthusiasm of a person trying to obtain a basic understanding of a particular science. Geology too is fraught with a plethora of terms that can be foreign to folks who would like to have a better understanding of their earthly environment. I have tried to minimize the use of such terms. My intent is to present concepts, nomenclature, principles, and theories with clear definitions and examples relevant to Tennessee. In order to achieve that goal, I have included a liberal number of photographs and illustrations to assist readers in comprehending geological concepts and recognizing locations of geologic significance across the state of Tennessee.

Earth science recently has been accepted to satisfy a required science unit for entrance into both the Tennessee Higher Education Commission and the Tennessee Board of Regents institutions of higher education. It is my hope that this book can be a resource for teachers and students to understand better the historical aspect of the earth sciences.

Currently the paradigm for teaching earth science is "Earth Systems Science." It employs the concept that the Earth consists of five interacting parts or systems (spheres), called the geosphere, hydrosphere, atmosphere, biosphere, and cryosphere. I have used the systems model to illustrate how the spheres interact in ways that affect our everyday lives and, more importantly, how they have interacted for billions of years to generate the rock record that forms the documents of Earth history.

Geologists using a forensic approach to reconstruct Earth history conclude that their interpretations of the past can be used to prognosticate the future of our planet. Also, readers will find that a hidden message relating to present global concerns threads through this book.

It is important to note that the endeavor of writing this book would not have been possible if not for the compilations of geologic data by the many geologists to whom I refer in chapter 5 as the "Giants of Tennessee Geology."

Acknowledgements

There are many individuals to whom the author is grateful for their roles in making this project a reality. First on this list are George D. Swingle and Harry J. Klepser. Although both are deceased, they were my mentors during graduate school at the University of Tennessee who instilled in me a love for Tennessee geology. Others in the Knoxville geologic community during my early years in Tennessee, but now deceased, to whom I give a great deal of credit for helping me better understand Tennessee geology include Stuart Maher, Robert Laurence, and Jack Kellberg. Also, Bob Milici, presently with the United States Geological Survey (USGS) needs to be acknowledged for the many occasions he shared his knowledge of the Appalachians.

Harry Moore, retired TDOT geologist and author, read iterations of my manuscript more than once, but most of all kept me plugging along with words of encouragement. Pete Lemiszki with the Tennessee Division of Geology kindly read the manuscript and made many helpful suggestions.

I am especially indebted to Michael A. Gibson, professor of geology at the University of Tennessee at Martin, for his meticulous review of the manuscript. His suggestions have greatly enriched the quality of the book and certainly helped me achieve the goals I hoped for.

Katherine Aycock was helpful in recruiting Linda Weaver who did a magnificent job of editing. Her limited background in geology kept me mindful of the reading audience that I wanted to target.

And, last but not least, I am indebted to Valerie Hunt, teacher at the Bowman School in Cleveland, Tennessee, for a very large portion of the artwork. Truly her drawings are worth millions of words.

Chronological Outline of Tennessee Geology

TIME FRAME	TIME UNITS	TECTONICS
~1800–1100 Ma	L. Paleoproterozoic–Mesoproterozoic Eras	Collision of continents–Grenville orogeny; Supercontinent Rodinia formed.
~750–734 Ma	Neoproterozoic Era	Breakup of Rodinia; Reelfoot Rift fails to separate proto–North America into two continents; rift basins form along the trend of the present day Appalachians; Iapetus Sea begins to form.
~600 Ma	Neoproterozoic Era	Melting of glaciers on other continents add to sea level rise.
~500Ma	Neoproterozoic Era	Sea floor spreading continues—Iapetus Sea spreads as eastern proto-North America becomes a Passive Margin. Sediment eroded from interior of continent (Craton).
~475 Ma	Early Ordovician Period	Iapetus Sea regresses from proto–North America exposing a terrain of carbonate. bedrock
~472 Ma	Middle Ordovician Period	Taconic orogeny causes uplifting and volcanic eruptions along eastern margin of proto– North America; a deep foreland basin is down-warped in the Appalachian region as shallow seas transgressed across Tennessee.
~444–385 Ma	Silurian and Devonian Periods	Mostly erosion in East Tennessee (deep into previously deposited middle Ordovician rocks); continued uplift (related to Taconic and Acadian Orogenies) in the East serves as sediment source; uplifting and down-warping in Middle and West Tennessee (Nashville and Ozark Domes) influence erosion and deposition in Central and West Tennessee; possible timing of Flynn Creek and Wells Creek impact structures.
~385–326 Ma	Late Devonian and Early Mississippian Epochs	A shallow sea transgressed across Tennessee but was gradually displaced by clastic sediment derived from uplifted areas in the east as various terranes continue to collide with proto-North America
~326–312 Ma	Late Mississippian–Middle Pennsylvanian Epochs	African plate begins colliding with proto-North America
~300–250 Ma	Carboniferous–Early Triassic Periods	Alleghanian orogeny; Africa collides with North America; Appalachian region (Blue Ridge, Valley and Ridge, and Cumberland Plateau) tectonically moved ~100 miles westward along a major decôllement

TENNESSEE ROCK RECORD

Cloudland Granite Gneiss and Cranberry Granite Gneiss exposed in northeastern Tennessee; occurs as crystalline basement beneath younger sedimentary veneer across Tennessee.
Intrusion of Beech Granite and Bakersville Gabbro in northeastern Tennessee; the Ocoee Supergroup (bedrock in the Great Smoky Mountains National Park) formed in the rift basins; volcanics and glacial deposits in Mount Rogers rift.
Deposition of the Sauk Sequence; best exposures are in East Tennessee—mostly covered in Middle and West Tennessee.
The end of the Sauk Sequence; karst develops on exposed carbonate rocks (Knox Group).
The Tippecanoe Sequence begins with deposition upon the karst erosional surface to create a paleokarst disconformity; layers of volcanic ash (bentonite) are deposited within the Chickamauga Group in East Tennessee, and the Stones River and Nashville Groups in Middle Tennessee; fossiliferous Tippecanoe rocks are exposed in nearly all parts of Tennessee except the Cumberland Plateau (covered by younger strata)—the Great Ordovician Biodiverity Event (GOBE).
A disconformity marks the end of the Tippecanoe Sequence as coarse clastic wedges of sediment shed from rising mountains were deposited in the East; in Middle and West Tennessee complex facies including the Brassfield Limestone, Wayne and Brownsport Groups, and Decatur Limestone were deposited west of the sediment prograding from the east; the Ross Formation (includes the fossiliferous Birdsong Shale) and the Camden Formation were deposited in the remaining water of the sea that was regressing westwardly. The last Tippecanoe formation deposited in the regressing sea was the Pegram Formation.
The Chattanooga Shale (black) of a transgressing sea marks the beginning of the Kaskaskian Sequence; fossiliferous Mississippian limestone was deposited in shallow, well-lit seas; crinoids (pelmatazoa), and coral are abundant; limestone grades upward into shale and other clastic rocks as sediment eroded from the uplifted areas in the east bringing a close to the Kaskaskian Sequence.
Absaroka Sequence begins ~318 Ma; deltas prograding from east to west form the coal-forming paleoenvironments (lagoons, bayous, swamps, etc.) for the rocks that now cap the Cumberland Plateau and Cumberland Mountains.
As Africa is subducted beneath North America the Blue Ridge Mountains are raised to Himalayan proportions; supercontinent Pangea is formed.

Chronological Outline of Tennessee Geology (cont.)

TIME FRAME	TIME UNITS	TECTONICS
~228 Ma– Present	Late Triassic Epoch–Present	Pangea breaks up similar to Rodinia (750 Ma); rift basins develop along Appalachian trend develop but none known in Tennessee.
~250–99.6 Ma	Late Triassic Epoch–Late Cretaceous Epoch	Erosion over most of Tennessee; Mississippi Embayment possibly passes over the Bermuda Hot Spot.
~99.6–65.5 Ma	Late Cretaceous Epoch	A Cretaceous sea transgressed and regressed in the Mississippi Embayment of the Gulf Coast in West Tennessee.
~65.5 Ma–Present	Paleogene–Neogene Periods	The K-T (or K-P) Boundary between the Cretaceous and Paleogene Periods noted as the time of dinosaur extinction; marine and nonmarine sedimentary deposits represent the last oceans that transgressed and regressed up the Mississippi Embayment of the Gulf Coast of West Tennessee.

TENNESSEE ROCK RECORD

	The Atlantic coast of North America became a passive margin; sediment from the ancient Appalachian Mountains is transported to the passive margin or a trough located at the New Madrid rift (Reelfoot Rift) in West Tennessee.
	Zuni Sequence: Tuscaloosa gravel lies disconformably upon older Devonian and Mississippian rocks (often filling ancient sinkholes to form a paleokarst); sedimentary facies include the famous fossiliferous Coon Creek Formation (home of the Tennessee State Fossil, *Pterotrigonia thoracica*.
	Tejas Sequence: deposition of marine and non-marine sediments similar to the Zuni Sequence; Gray fossil site in East Tennessee; loess deposits formed from glacial outwash in the Mississippi River Valley; balds and block fields due to periglacial climate in the Blue Ridge Mountains.

The Geologic Setting of Tennessee

> Owing, for the most part, to the extent and varied nature
> of the country traversed by the State, one of its most promi-
> nent characteristics with reference to natural Features, is
> great variety.
>
> —Dr. James M. Safford, *Geology of Tennessee* (1869)

Why is Tennessee divided into three grand divisions? It might be said that politics distinguish East, Middle, and West Tennessee. Although that may be in part
true, there is, perhaps, something that lies deeper, some reason for any perceived
political differences. Tennessee is a long and narrow state—extending from east
to west approximately 440 miles between 81° 37′ West longitude and 90° 28′
West longitude and extending from south to north about 120 miles between 35°
North latitude and 36° 41′ North latitude. Each grand division has a manifestly
different landscape resulting from erosion of distinctly different rock types, rock
structures, and climate. These differences are the result of both past and present interactions of the Earth's geosphere, hydrosphere, atmosphere, biosphere,
and cryosphere. Geographically speaking, regions having similar landforms are
recognized as **physiographic provinces** (Fenneman 1928, 1931, and 1938). One
or more distinct physiographic provinces embrace each of Tennessee's grand divisions. Figure 1.1 is a **block diagram** (a drawing that shows three dimensions
by combining a land surface map view with two cross-sections) illustrating the
major geologic structures and the physiographic divisions of Tennessee.

The mountainous terrain of the Eastern Grand Division begins with the Blue
Ridge province, or Unaka Mountains, which contains the state's highest elevation—Clingmans Dome, at 6643 feet above sea level—and extends westward to
include the Valley and Ridge province and Appalachian Plateaus province (en-
compassing the Cumberland Plateau and Cumberland Mountains). Middle Tennessee is part of the Interior Low Plateaus province, which includes the Nashville
Basin and the Eastern and Western Highland Rims. The Western Division is part
of the Gulf Coastal Plain province, which includes the Western Valley of the Tennessee River, the West Tennessee Uplands, and the lowest elevation in the state—

the Mississippi River valley—at 178 feet above sea level. The mean elevation of Tennessee is 900 feet above sea level.

Although more detailed descriptions of the rocks and structures in the physiographic provinces will be presented in subsequent chapters, a brief overview of the landscapes in each of the provinces is presented here.

The Blue Ridge Province

That part of the Blue Ridge, or Unaka, Mountains in Tennessee begins at the North Carolina–Tennessee state line and extends westward to a fault line that separates old, highly deformed metamorphic rocks (see chapter 8) that are very mountainous from younger and less deformed sedimentary rocks of the Valley and Ridge province (see chapter 9), which consists of alternating ridges and valleys.

Great Smoky Mountains National Park (an International Biospheric Reserve) is the centerpiece of the province; there are many spectacular mountain summits and waterfalls to behold. Several significant peaks over 5000 feet in el-

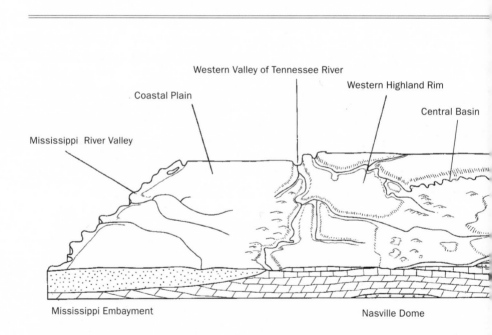

Mississippi River Valley

Coastal Plain

Western Valley of Tennessee River

Western Highland Rim

Central Basin

Mississippi Embayment

Nasville Dome

| Tertiary and Cretaceous sand | Mississippian, Devonian, Silurian, Ordovician, and Cambrian limestone | Ordovician and Cambrian dolomite |

evation include Clingmans Dome, which has an accessible observation tower, Mount LeConte (see figure 1.2), Thunderhead Mountain, and Mount Kephart. Other summits include grassy balds (treeless summits with grass and/or shrub vegetation) such as Gregory Bald, Andrews Bald, and Silers Bald, all offering spectacular vistas as well as springtime displays of flame azaleas (see figure 1.3). Waterfalls and cascades abound, but a few favorites are Ramsey Cascades, Laurel Falls, Abrams Falls, Rainbow Falls, and White Oak Sink (see figs. 1.4 and 1.5).

All the rocks making up the Blue Ridge originated 100 miles or more east of their present positions. The rocks were intensely deformed and pushed west during a mountain building event when proto–North America collided with another continent (see chapter 13). During this event, the older Blue Ridge rock layers

Figure 1.1. Block diagram of Tennessee showing Tennessee's physiographic provinces and the relationships of rock types and structure to the landscape. The eastern grand division embraces the Unaka Mountains, or Blue Ridge, the middle division extends from the Cumberland Plateau to the Western Valley of the Tennessee River, and the west division includes the Coastal Plain and the Mississippi River Valley. (Miller 1974.)

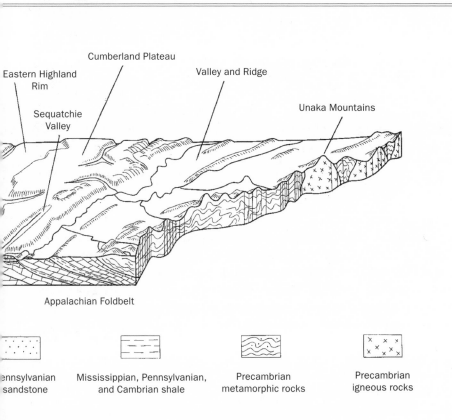

Cumberland Plateau

Eastern Highland Rim

Valley and Ridge

Sequatchie Valley

Unaka Mountains

Appalachian Foldbelt

Pennsylvanian sandstone

Mississippian, Pennsylvanian, and Cambrian shale

Precambrian metamorphic rocks

Precambrian igneous rocks

Figure 1.2. A view of Mt. LeConte from Clingmans Dome. Note the scars on LeConte's slope. They are rockslides that are part of the mass-wasting process gradually wearing away the mountain.

Figure 1.3. A view looking east from Andrews Bald, Great Smoky Mountains National Park. Andrews is one of the balds in the Smokies that is managed (trees and scrubs cleared) in order to maintain its condition as it existed at the close of the ice age and then later maintained for grazing purposes.

were thrust or stacked on top of younger rock layers as **thrust faults**. Features in the Blue Ridge providing evidence of this westward movement are the low pastoral areas, called coves, surrounded by mountains. The coves represent geologic windows where erosion has cut downward through the westerly transported old rocks into the younger rocks below. The rocks on the ridges surrounding the valley floor are older than the rocks forming the valley floor, indicating that a fault line surrounds the cove like a ring around the inside of a bathtub. Above the line are old rock layers (very resistant to erosion), and below the line, forming the coves, are the younger, more easily eroded rock layers (mostly limestone and dolomite). Cades Cove, Tuckaleechee Cove, and Wear Cove are good examples of geologic windows (see figure 1.6).

Figure 1.4. Left, Ramsey Cascades, Great Smoky Mountains National Park. One of many waterfalls and cascades formed by streams flowing over ledges of resistant Precambrian rocks.

Figure 1.5. Above, White Oak Sink, Great Smoky Mountains National Park, along the southern border of Tuckaleechee Cove. Water cascades over resistant, older Precambrian rocks that have been thrust faulted over younger, soluble Paleozoic limestone into a cave system that flows to Tuckaleechee Cove. Both have been eroded through, resulting in the cave system below the waterfall.

The Valley and Ridge Province

The province is so named for the obvious alternating elongate northeast-southwest oriented valleys and ridges. It extends diagonally across the width of East Tennessee bounded by the foothills of the Blue Ridge on the east and the steep escarpment of the Cumberland Plateau on the west (see figure 1.1). The layers of sedimentary rocks forming the province are dominantly tilted to the southeast as a result of the same forces that deformed the rocks of the adjacent Blue Ridge. Those layers of tilted rocks—composed of earth materials resistant to erosion, such as sandstone and dolomite containing abundant chert form the ridges—and the intervening valleys are underlain by rocks less resistant to erosion, such as limestone or shale. The stacking of the rock layers one upon another, in a fashion resembling shingles on a roof, and separated by thrust faults (rock ruptures) causes sequences of layers to be repeated across the province. Figure 1.7 is a

Figure 1.6. Cades Cove, Great Smoky Mountains National Park. The floor of the cove is underlain by limestone of Ordovician age, and the surrounding mountains are much older rocks of Precambrian age that were thrust faulted onto the younger limestone.

generalized cross-section illustrating the relationship between the rocks in East Tennessee and the topography.

Some rock units tend to form characteristic shaped ridges and valleys or favor certain types of vegetation. For example, the Rome Formation, typically underlain by a fault (see figure 1.7), forms a narrow cockscomb-shaped ridge when viewed from the side and supports mostly pine trees (see figure 1.8), whereas cherty dolomite of the Knox Group typically forms a rather broad ridge with oak and hickory trees. By contrast, limestone valleys generally support juniper (cedar) trees. Some ridges are prominent enough that they are referred to as mountains. Some of the more prominent ridges include Powell, Clinch, and Bays Mountains in the northern portion of the province and Whiteoak Mountain in the southern portion. Most ridge elevations are between 1000 and 2000 feet, but range up to just over 3000 feet for Bays Mountain. Valley elevations range from around 1000 feet in the northern province to about 800 feet in the southern portion.

The Cumberland Plateau

The western boundary of the Valley and Ridge province is marked by an abrupt escarpment that rises to a relatively flat upland called the Cumberland Plateau, recognized as a section of the more extensive Appalachian Plateaus province of the Appalachian Highlands (see figure 1.9). The height of the escarpment averages 900 feet. Because the rocks under the plateau are less deformed than the two provinces to the east, the plateau is capped by nearly flat-lying sedimentary strata. South of Interstate Highway 40 (I-40) the plateau is more truly flat,

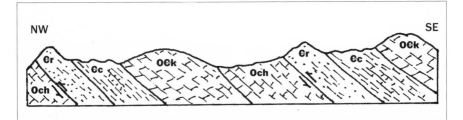

Figure 1.7. A generalized cross-section of a small portion of the Valley and Ridge province showing the relationship between rock formations and topography. From oldest to youngest the rock units are Єr = Rome Formation, Єc = Conasauga Group, OЄk = Knox Group, and Och = Chickamauga Group. The arrows show relative movement along thrust faults. Note also that the stratigraphic section is repeated as a result of faulting. The drawing is not to scale. Rock units are discussed in more detail in chapter 7.

Figure 1.8. Bull Run Ridge underlain by the Rome Formation. The hummocky "cock's comb" topographic profile is typical of ridges underlain by the Rome and generally designates the location of a thrust fault in the Valley and Ridge province. The view is from Raccoon Valley looking south.

with elevations averaging 1700 to 1900 feet, but north of I-40 is a section often referred to as the Cumberland Mountains, where the terrain is higher and more mountainous, with elevations reaching over 3000 feet. The highest elevation in the Cumberland Mountain section is Cross Mountain (3534 feet) near Lake City.

Although the rocks of the plateau are considerably less deformed than those of the Valley and Ridge and the Blue Ridge provinces, several features lend evidence that deformation did affect the region. The "Devil's Racetrack" near Cove Lake State Park, and viewable from I-75, is a spectacular exposure of vertically oriented layers of rock that were rotated into that position from their original horizontal position by mountain-building forces (see figure 1.10). Also, two elongate valleys carved by erosion into the plateau, Sequatchie Valley south of Crossville (see figure 1.1) and Elk Valley in Campbell County, are related to deformation.

Figure 1.9. A view of Lookout Mountain, Chattanooga, showing the steep escarpment of the plateau. Note the sharp cliff at the summit formed by resistant sandstone and the concave slope below underlain by less resistant shale. (Photograph by Jim Hunt.)

Figure 1.10. Devil's Racetrack, looking northeast. The near vertical layers of Pennsylvanian sandstone are part of the western flank of the Powell Valley anticline and were no doubt influenced by movement along the Jacksboro fault located in the adjacent valley to the southwest. The layers flatten to being almost horizontal a short distance to the northwest. The sandstone is more resistant to erosion than the shale between the pinnacles.

More easily eroded limestone (often riddled with caves) occurs beneath the resistant sandstone capping the plateau, and as the limestone gradually erodes, the edge of the plateau retreats, reducing the size of the plateau. Whereas the eastern edge of the plateau is a relatively well-defined escarpment, the western boundary is quite irregular. This is attributed to the fact that the drainage divide for the plateau is nearer to the eastern edge than it is to the western edge, causing westward-flowing streams to erode rather deep gorges that extend nearly to the eastern edge of the plateau. In these gorges, the erosion-resistant sandstone that caps the plateau forms many spectacular waterfalls. Fall Creek Falls, located in the state park by that name, is perhaps the most famous. At 256 feet, it is considered one of the highest in the eastern United States (see figure 1.11).

Figure 1.11. Fall Creek Falls. The waterfall cascades over a resistant caprock of Pennsylvanian age sandstone. Sandstone layers such as this form the resistant cap of the Cumberland Plateau.

Within the limestone, below the plateau's sandstone cap along the western edge of the plateau, is a rock formation with a high silica content resistant enough to erosion that it forms a cap above the underlying, more easily eroded limestone. This creates a relatively flat area below the plateau called the Highland Rim (see figure 1.1). The Highland Rim marks the western boundary of the Cumberland Plateau.

The Highland Rim

According to Fenneman (1938), the Highland Rim is considered part of the Interior Low Plateaus province that also includes the Central Basin of Tennessee (see figure 1.1). The Highland Rim surrounds the Nashville Basin, making it divisible into the Eastern Rim, which borders the Cumberland Plateau to the east, and the Western Rim, which adjoins the Western Valley of the Tennessee River on the west. Even though the Eastern Rim averages only 25 miles in width, it is distinct. Traveling east to west from the Cumberland Plateau to the Eastern Highland Rim and across the Highland Rim to the Nashville Basin (a topographic basin as opposed to a structural basin) there is a rather marked stair-step drop in elevation of several hundred feet between the rim and the Cumberland Plateau on the east and the rim and the Nashville Basin on the west. In general, topography of the Eastern Rim is flat, but it is also characterized by features of limestone erosion, such as sinkholes and caves. Many of the gorges described along the western edge of the Cumberland Plateau extend as narrow valleys across the rim into the basin. A few erosional remnants of the Cumberland Plateau, capped with the same rocks that cap the plateau, remain on the rim; the most notable is Short Mountain in Cannon County, with an elevation of 2074 feet and located about 20 miles west of the western edge of the plateau.

The land surface of the Western Rim is less well defined than the nearly flat Eastern Rim. The Western Rim is highly dissected by drainage, creating a rolling landscape. The Western Valley of the Tennessee River forms the western boundary of the Highland Rim (Miller 1974).

The Central or Nashville Basin

The Central Basin, surrounded by the Highland Rim, is definitely a topographic basin when compared to the surrounding uplands. But, interestingly enough, the Central Basin is, in structural sense, a dome. During rock deformation in the geologic past, the central part of Tennessee was actually uplifted, forming the Nashville Dome (see figure 1.1). During the doming, erosion resistant rocks were ultimately breached, and more easily eroded limestone was exposed. Percolating groundwater gradually dissolved and removed the limestone creating the topographic basin. The basin is relatively flat in the center and somewhat hilly around the outer basin where erosional remnants of the Highland Rim remain.

The Western Valley

The Tennessee River flows north across West Tennessee as it returns after leaving East Tennessee to flow across Alabama. The history of the peculiar course of the Tennessee River is puzzling. Chapter 15 discusses possible explanations for its development. The valley, on the average, is 20 miles wide, and the floodplain ranges from 1.5 miles to 3.5 miles in width (Miller 1974).

The Coastal Plain

The Coastal Plain represents an embayment of a former ocean that extended in the geologic past from the Gulf of Mexico to about the present-day location of Cairo, Illinois. For that reason, it has most of the same attributes as the coastal areas of Alabama, Mississippi, and Louisiana. Except for some hilly terrain that forms the divide between the Tennessee and Mississippi Rivers, the land is flat and slopes gently toward the Mississippi River Valley. The stratigraphic units of Coastal Plain are mostly unconsolidated sediment and account for some of the youngest chapters in Tennessee's geologic history.

Mississippi River Valley

The valley of the Mississippi River forms the western border of Tennessee with Missouri and Arkansas. The nearly 15-mile wide floodplain of the meandering Mississippi contains all the typical geologic elements—oxbow lakes, natural levees, swamps, and meander cutoffs. Possibly the most significant geologic feature of the valley is Reelfoot Lake in Lake County (fig. 1.12). The origin of the lake is related to the New Madrid earthquakes that shook the region for nearly a year during the period 1811–1812.

Figure 1.12. Reelfoot Lake. The lake formed when the land subsided during the New Madrid earthquake (1811–1812). Many upright trees remain submerged below the lake, and the cypress trees above the water were rooted on natural levees of a pre-earthquake stream.

2

A Dynamic Planet

> There hasn't been a moment when I had the chance to look
> down on our planet from orbit when I haven't been amazed
> at how geology has played a significant role in the develop-
> ment of humankind.
>
> —Dr. James F. Reilly Jr., NASA astronaut/geologist

The Earth's history is so vast compared to a single human life span that it is rather difficult to comprehend the effects that interaction among the Earth's systems can produce. Current calculations indicate that the Earth formed at least 4.5 billion years ago. What that means with regard to Tennessee's geologic history is that nearly three-fourths of Earth's history had already transpired before the history covered in this book occurred. Although the geologic history described in this book began as early as 1.2 billion years ago, most of Tennessee's geologic record spans only the last 800 million years. Over the course of a billion-year span, what we presently recognize as Tennessee's landscape would have changed many times. In the past, parts of Tennessee might have resembled the present-day Bahamas, the South Pacific, the Himalayan Mountains, the Carolina coasts, or the Mississippi Delta. Before embarking on a journey to understand how these different environments came to pass in the annals of Tennessee history, it is necessary to understand a few significant facts about planet Earth.

The first and perhaps most important fact is that the Earth is a dynamic planet. It is a natural system consisting of a set of subsystems that have been interacting ever since the "blessed event"—the origin of the Earth. These subsystems are the **geosphere**, **hydrosphere**, **atmosphere**, **biosphere**, and **cryosphere**. Their interactions have generated the rock records, preserved as part of the geosphere, that are used to decipher Earth history. Figure 2.1 depicts the interrelationships among the five subsystems.

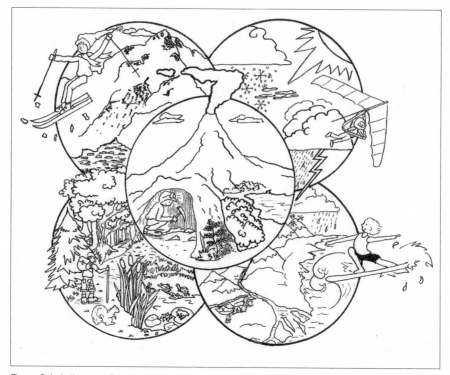

Figure 2.1. A diagram of the five subsystems of the Earth System showing how they are interconnected. The center sphere represents the geosphere that interacts with all of the other four. Clockwise from the upper right are the other interacting spheres, the atmosphere, the hydrosphere, the biosphere, and the cryosphere.

Geosphere

The geosphere embraces earth materials from the Earth's center to its surface. Physical, chemical, and thermal processes within the Earth can modify the properties of earth materials and generate forces capable of moving the outermost layer of the Earth, causing earthquakes and volcanoes. The moving pieces on the Earth's surface are called **plates**, and **plate tectonics** is an important theory that unifies our understanding of the evolution of the geosphere. The moving plates have profoundly influenced the last billion years of Tennessee geologic history.

The geosphere can be compared to the concentric layers of an onion (see figure 2.2). An **inner core**, about 1500 miles (2462 km) in diameter, is composed of solid iron and nickel. Around the inner core is a liquid **outer core**, about 1400 miles (2257 km) thick. The core is in turn surrounded by the **mantle**, a layer of minerals called iron-magnesium silicates that can be divided into upper and lower mantles. The lower mantle is hard, solid, and approximately 1575 miles

(2533 km) thick, while the upper mantle, called the **asthenosphere**, is a soft solid about 155 miles (250 km) thick. The outermost 60 miles (100 km) above the asthenosphere is the more rigid **lithosphere** that contains an outer rind called the **crust.** Oceanic crust ranges from 3 to 5 miles (5 to 8 km) in thickness, and continental crust ranges from 16 to 45 miles (25 to 70 km) in thickness. The hard lithosphere gliding across the soft asthenosphere constitutes plate tectonics. Internal changes of the geosphere create obvious external effects on the Earth's surface in the form of volcanoes, earthquakes, and mountain belts, but also generate the forces that drive plates of lithosphere across the mantle.

Earth's lithosphere might be compared to a cracked eggshell, with the cracks separating the pieces of shell and simulating plate tectonics. As pieces of eggshell spread apart along a crack, other pieces of shell necessarily converge and/or overlap along other cracks. Since the size of the Earth has remained essentially the same throughout most of geologic time, the cracked eggshell models the kind of process at work in the plates. High temperatures within the Earth generate convection currents in the puttylike plastic mantle, bringing hot, soft mantle toward the Earth's surface and causing plates of lithosphere to spread apart, forming **rifts** (low areas formed by plates moving apart).

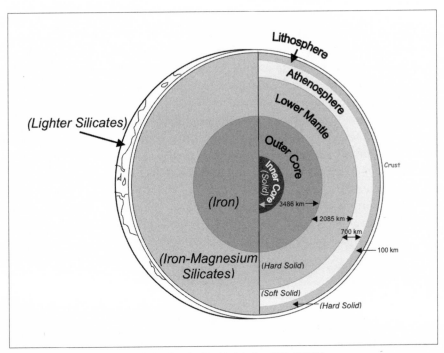

Figure 2.2. A generalized model showing the Earth's interior. (Lillie 2005.)

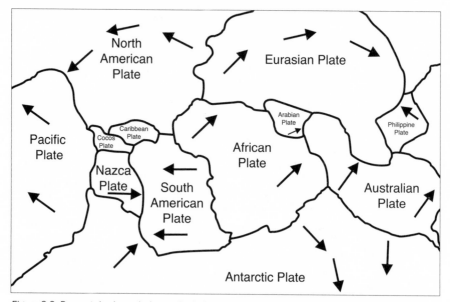

Figure 2.3. Present-day boundaries and relative motions of the Earth's plates. The western edge of the North American Plate is an example of a convergent boundary whereas the eastern edge is an example of a passive margin of a divergent boundary.

Just as in the eggshell model, the boundaries of plates on Earth can be associated with **divergence** (spreading apart or rifting) or **convergence** (collision or subduction). Rates of plate movement are extremely slow, measured to be a few centimeters per year or, by analogy, about the rate of fingernail growth. Such a rate may at first appear insignificant, but when viewed in the context of **deep time** involving billions of years, it becomes significant. For example, at the present rate of plate movement, an unimpeded drifting plate at the equator could be shoved around the circumference of the Earth just short of two times during a billion years. Figure 2.3 shows the current position of Earth's plates and their relative motions.

Large portions of plates above sea level form the continents of the world; smaller pieces of "real estate" drifting about on plates are referred to as **terranes**. In the geologic past many plates collided and were sutured together to form a supercontinent.

Hydrosphere

When compared to all the other planets in the Solar System, Earth could aptly be referred to as the water planet. The hydrosphere includes all the liquid water on the Earth, including the oceans, rivers, streams, and subsurface water. The moniker water planet is obvious when the planet is viewed from space—nearly three-fourths of the Earth's surface is covered by water, with the greatest expanse being ocean. The geographic relationship between oceans and landmasses has changed many times during the last billion years. The study of these ancient relationships is termed **paleogeography**. There were times when what is now land-locked Tennessee periodically was inundated by shallow seas that transgressed and regressed across the continent. Tens of thousands of feet of sediment along with the remains of critters that were living during the incursions of the sea ultimately formed the fossiliferous sedimentary rocks that record a large portion of the geologic history of Tennessee.

During those times when Tennessee was above sea level, fossils of terrestrial life and sediment deposited by running water of streams and rivers contributed to the geologic records. The coal-bearing rocks capping the Appalachian Plateau (Cumberland Plateau) were formed from sediment deposited in meandering rivers, lakes, marshes, barrier beaches, and bayous as parts of ancient deltas, similar to the present Mississippi Delta. Erosion takes time. The diverse landscapes of ridges, valleys, plateaus, and mountains across Tennessee were sculpted by running water during the last 250 million years or so. The present landscape is still being modified by erosion, with added impact from the modern biosphere (humans) seeming to be the most profound. All in all the geologic history of Tennessee is like a history book that has had some pages and even chapters torn out. The pages representing ancient marine environments are dominantly depositional and better preserved, but those portions of the terrestrial record tending to be dominated by erosion and sediment removal are the pages or chapters that have been removed. However, when all the processes are considered, the sum total of geologic time in Tennessee is represented.

Water in different ways has been the major agent recording geologic history, either writing pages of history (deposition) or removing them (erosion). It should go without saying that the hydrosphere is one of the Earth's most important systems, especially with regard to life on Earth. Water is the planet's most important natural resource. Without it, life as we know it would be impossible.

Atmosphere

Direct evidence of atmospheric changes in the geologic record over the last billion years is sparse. However, because of the relationship between the biosphere and the atmosphere, what is known about conditions of past atmospheric conditions or climate changes are deciphered from indirect evidence found in the fossil record. By invoking eighteenth century geologist James Hutton's principle of **Uniformitarianism** (the present is the key to the past), fossil plants (including pollen and spores) and animals can provide clues about temperature, composition, and movements of the atmosphere in the geologic past. Just as present-day ecosystems comprise the relationships of organisms with their environments, relationships between fossils and ancient atmospheres (or climates) can be interpreted. **Paleoecology** is the study of these relationships—more specifically, **paleoclimatology,** which, in addition to the fossil record, draws on other lines of evidence, such as studies of carbon and oxygen isotopes preserved in glacial ice and tree rings.

Based on the fossil record, the rise in levels of atmospheric oxygen to today's level is equated with the explosion of land plants and photosynthesis. Also the relative abundance of iron oxide associated with sedimentary rocks is suggestive of significant levels of atmospheric oxygen. Fossil pollen in lakebed sediments deposited during ice ages has been used to reconstruct past climatic conditions. Pollen from tree species associated with a temperate climate is indicative of warmer interglacial ages, or a location far away from an ice sheet, while pollen from boreal tree species indicates a location proximal to an ice sheet. Also, another biosphere-atmosphere relationship indicating times past when the climate was likely to have been more tropical than that of today is found in comparing environments where modern coral reefs are forming today to the coral fossils found in some Tennessee rocks.

Biosphere

The biosphere includes all living matter, including us. It should be very obvious that humans are active in Earth processes. Human history is full of accounts of cultures that have "chosen," so to speak, to "fail" through their adverse impacts on the environment (Diamond 2005). Although the present trend toward global warming is part of Earth's response to systemic interactions, human activity (the biosphere) has no doubt contributed to the warming. Examples of **anthropogenic** changes to the landscape are numerous. In fact, human activities can actually accelerate rates of geologic processes such as erosion. With explosives, bulldozers, chainsaws, and warfare, certain Earth systems phenomena that normally require millennia to occur can be accomplished in a matter of seconds.

Organisms play essential roles as architects in the formation of certain sedimentary rocks. For example, most carbonate rocks, such as limestone and dolomite, are biochemical products—i.e., fossil reefs and coquina limestone. Also, consider that most of the sediment in fine-grained sedimentary rocks such as shale is undigested material that once passed through alimentary canals of millions of filter-feeding animals—actually **coprolites** (fossil feces).

Cryosphere

The cryosphere embraces all the solid water, or ice, on the planet—ice caps, continental ice sheets, and glaciers. Direct evidence of glaciation in Tennessee is scant. The only direct evidence is found in rocks about 700 million years old (Neoproterozoic; see chapter 4 on geologic time), found in the extreme northeastern corner of Tennessee in proximity to Mount Rogers, Virginia. However, there is indirect evidence of cryospheric impact on the geologic history of Tennessee. For example, much of the sediment in the Mississippi River Valley of West Tennessee is outwash from glaciers that periodically pushed southward to near the present Tennessee-Kentucky border 1 million to 11,000 years ago. The prominent vertical bluffs overlooking the Mississippi River are composed of **löess**, a fine, wind-blown sediment deflated from the glacial outwash by westerly winds. Another possible indirect effect of glaciers is sea-level fluctuations caused by the waxing and waning of glaciers on various parts of the planet. Such fluctuations of sea level can possibly be equated with some of the transgressions and regressions of the sea across the continent between 542 and 250 million years ago.

To further illustrate how Earth's subsystems can interact in today's world, consider past eruptions of volcanoes (geosphere) in the Yellowstone region of the western U.S. when volcanic ash was transported by westerly winds (the atmosphere) many miles to the east. As the cloud of ash moved eastward, plants and animals (the biosphere) become buried in ash in a fashion similar to the classic example of Pompeii, Italy, or Ashfall, Nebraska (see figure 2.4). In the case of the eruption of Mount St. Helens in 1980, streams and rivers (the hydrosphere) became flooded through the influx of melt water from ice fields or glaciers (the cryosphere) that capped the once dormant volcano, and then the waters became choked with tons of sediment. Another scenario could be the biosphere (for example, us) far from the site of the eruption (geosphere) could be impacted economically due to destruction of crops by volcanic ejecta.

The process of understanding geologic processes has a long history—one could call it the history of geologic history. A major breakthrough in the understanding of geologic history can be related to James Hutton's basic principle of geology referred to earlier. It might be said that his principle of uniformitarianism, that "the present is the key to the past," was the beginning of modern geology. Although Hutton did not reference specifically the five subsystems of the Earth system, he recognized that Earth's natural processes have functioned throughout time, providing the records of geologic history. Modern environments such

Figure 2.4. One of many rhinoceros fossils preserved in volcanic ash at Ashfall, eastern Nebraska. The ash source has been traced to a location in southern Idaho. The organisms were grazing near a watering hole prior to being buried in ash about 12 Ma. Similar fauna are found hundreds of miles away in Tennessee at the Gray Fossil Site. Although the animals at that site are about the same age, they are entombed within sediment in an ancient sinkhole.

as the Bahamas and the Mississippi Delta are truly the types of analogs used for interpreting the rock records formed over the past hundreds of millions of years. Changes in Earth's subsystems can cause instantaneous changes such as landslides, volcanoes, and tsunamis in landscapes, but the rate of evolution of landscapes of the geologic past (e.g., the formation of mountains, oceans, deltas, etc.) wrought by these subsystems is exceedingly slow. For example, it is difficult to imagine the Colorado River eroding the Grand Canyon. However, the recognition of deep time, or the fact that Earth history spans billions of years, makes the acceptance of these changes credible. Just as in any study of history, one must be able to read the "language" in which the history is recorded. For Earth history this means: What can rocks tell us? The constant in the equation for the geologic history of Tennessee is the bedrock of the state. The variables in the equation are the "tools" for making observations and the perceptions of the observers.

3

Building A Record: Earth Materials and Processes

> Go my sons, burn your books and buy stout shoes, climb
> the mountains, search the valleys, deserts, the sea shores,
> and the deep recesses of the Earth.... Observe and experi-
> ment without ceasing, for it is in this way and no other will
> you arrive at a knowledge of the true nature of things.
>
> —Petrus Severininus, sixteenth-century Danish alchemist

Rocks and Minerals

The geosphere consists of earth materials that in turn are made of one or more minerals. Rocks and minerals not only provide natural resources that human-kind has learned to use to advance society, but they form the pages in the Earth's history book. A **mineral** is a naturally occurring inorganic chemical element or compound that has a characteristic atomic arrangement (**crystalline**) and physical and chemical properties that are fixed or varied within a given range. Nearly 5000 mineral species exist, but only 50 or so of those are considered common rock-forming minerals. Many good books are available for the amateur and the professional on how to identify minerals, but one simple way to identify minerals is by observing their physical properties. A list of dominant rock-forming minerals likely to be encountered in Tennessee, excluding those of economic importance, includes quartz, feldspar, mica, calcite, dolomite, and a few dark-colored iron-rich minerals.

A **rock** is an aggregate of one or more minerals or a partially or wholly glassy solid, which, had it formed under proper conditions, would have yielded a mineral aggregate. Rocks can be readily classified by using one or more of the following criteria: mode of origin, texture, or composition (mineral or chemical content). The three major classes of rocks, based on mode of origin, are **igneous**, **sedimentary**, and **metamorphic**.

Minerals

Because rocks are aggregates of minerals, in order to identify rocks it is important to be able to recognize some of the more common rock-forming minerals. The simple way to identify minerals is by determining a few characteristic physical properties and then to compare the result with information contained in an identification table. Although all materials have many physical properties, a few dominant physical properties identify most minerals. These properties are hardness, cleavage, and luster.

Hardness is a mineral's resistance to abrasion ("scratchability"). All minerals have been assigned a range of hardness according to the Mohs Hardness Scale, which is relative and ranges from one to ten, with diamond, the hardest known natural material, being number ten. Several common items useful for comparative testing of mineral hardness are a fingernail (2), a penny (~3.5), and a steel file, nail, or piece of glass (~5). For example, when testing a mineral, if it cannot be scratched by a penny but can be scratched by a nail or pocketknife, it is likely that the hardness of the mineral is 4.

Cleavage is the property whereby certain minerals break apart to produce smooth, well-defined surfaces. If a mineral has cleavage, it will split repeatedly along parallel planes. In the case of some minerals, the diagnostic feature will be their having multiple directions of cleavage. For example, a mineral with three perfect directions of cleavage would form cubes like the mineral halite (table salt).

The third useful property is **luster**, the appearance of the mineral in reflected light. Adjectives such as *metallic, greasy, glassy, earthy,* and *waxy* are typically used to describe luster. This is entirely different from a mineral's color. The color of a mineral is probably the least reliable physical property for identification; however, there are some minerals for which it is a diagnostic property.

Most mineral classification schemes group minerals according to their chemical composition. For example, there are oxides (elements bonded with oxygen); sulfides (elements bonded with sulfur); silicates (elements bonded with silicon and oxygen); and carbonates (elements bonded with carbon and oxygen).

Quartz, by definition, is an oxide of silicon (SiO_2), but it is generally regarded as a silicate mineral (see figure 3.1). It is abundant in rocks of all three classes and is the stuff we wiggle our toes in on most beaches, find in streambeds, or dig in the soil. The hardness of quartz is 7, and its luster is best described as greasy, but may also be glassy. It may occur in almost any color; one familiar example is purple amethyst. It should be noted that quartz does not have cleavage.

Feldspar is also a silicate mineral and occurs in all three classes of rocks (see figure 3.1). There are several varieties of feldspar, differentiated by composition. Basically, it is an alumino-silicate, meaning that it primarily contains the elements aluminum, silicon, and oxygen, but it also contains combinations of the

Figure 3.1. Four of the most common rock-forming minerals: 1. quartz (a clear crystal form), 2. feldspar (two prominent directions of cleavage), 3. biotite mica (one direction of cleavage), and 4. calcite (the rhombic shape with three directions of cleavage).

elements potassium, calcium, or sodium. Feldspar hardness is ~ 6, and its two-directional cleavage is a distinctive property. When observing rocks containing quartz and feldspar with similar color, the reflected light from the smooth flat cleavage of feldspar can help distinguish it from quartz, which does not have cleavage.

Mica, also a silicate mineral, has two common varieties—muscovite, which is clear, and biotite, which is black. Both varieties may be found in all three classes of rocks. Elements that occur in various amounts in different types of mica include potassium, magnesium, iron, aluminum, and fluorine. The characteristic properties of mica are that it is soft and easily scratched with a fingernail and that it has one perfect direction of cleavage, which provides its most distinctive property—flaking or splitting into very thin sheets (see figure 3.1).

Calcite is calcium carbonate. It has a hardness of 3 and three perfect directions of cleavage. When broken, the three directions of cleavage form perfect rhombohedrons (see figure 3.1). A distinctive test for calcite is that when cold, dilute hydrochloric acid (or vinegar in some instances) is dripped on it, the calcite effervesces or fizzes as carbon dioxide gas is liberated. Calcite is a mineral primarily found in sedimentary rocks, especially limestone, the official State Rock of Tennessee (so designated because of its abundance). It may be directly precipitated from seawater to form sediment or formed as dripstone deposits (stalactites or stalagmites) in caves. Most animals dwelling in the sea extract calcium carbonate from the water to form shells that may accumulate on the seafloor as sediment when the animals die.

Dolomite is another calcium carbonate mineral, but it also contains magnesium. The hardness is about the same as calcite, and it has three directions of cleavage. The surface of one cleavage direction is subtly convex or concave (saddle-shaped depending on which side is viewed). The saddle shape is helpful in distinguishing dolomite from calcite, but more commonly the cold, dilute hydrochloric acid test is used. Dolomite only mildly effervesces (sometimes only after being scratched), while calcite effervesces rather vigorously without scratching. Like calcite, dolomite is a mineral primarily found in sedimentary rocks. Other important minerals that are much less abundant, but are significant natural resources, will be discussed in chapter 16, which deals with economic geology.

Rocks

Of the three classes of rocks, igneous, sedimentary, and metamorphic, it is especially the sedimentary rocks that can be considered the chief documents of Earth history. However, every rock has a story to tell. It becomes a forensic activity to explore how, why, where, and when a rock was formed. Threading together the stories rocks have to tell constitutes Earth history. Unfortunately there is no one spot on the Earth that contains the complete story from the "blessed event," the birth of the Earth, to the present time. Some locations, however, such as the Grand Canyon in Arizona, contain several volumes of Earth history, if not the total story. It is necessary to piece together bits of "pages and chapters" from many locations around the world to get as near as possible to a complete geologic history of the Earth. The oldest rock records, or "chapters," in the geologic history of Tennessee occur in the eastern part of the state, while the most recent history is recorded in the sedimentary deposits in the western part.

The class of rocks called igneous all formed from the crystallization of molten material generated in the interior of the Earth. This molten material while within the Earth is called **magma;** however, once it is ejected onto the Earth's surface it is referred to as **lava.** Magma, when it solidifies within the Earth, forms intrusive bodies of igneous rock called **plutons.** When tabular in shape with boundaries parallel to layering of the intruded rock, the pluton is **concordant** and called a **sill.** Tabular plutons with the boundaries or contacts cross-cutting the layering of the intruded rock are **discordant** and called **dike.** Other than by drilling into them, plutons are observable only after the rock above them has been eroded away. Lava is extrusive—most commonly associated with volcanoes or lava flows. Igneous rocks are readily classified on the basis of their chemical or mineral composition and texture (the size of the minerals in the rock). Given slower rates of cooling and solidifying, plutons are typically coarser-grained than rocks formed from lava that cooled more rapidly on the surface of the Earth. The percentage of silica (SiO_2, mostly in the form of quartz) in the mineral composition of a rock is an important criterion frequently used in the classification of igneous rocks. One way to equate the amount of silica in a rock is to determine the amount of quartz present—the more siliceous rocks, such as granite, will normally contain higher percentages of quartz. Another observation is that, as the percentage of free silica as in quartz decreases, the iron and magnesium that impart darker colors to rocks becomes more abundant in the silicate minerals (minerals such as hornblende and augite). Figure 3.2 is a simplified scheme for classifying igneous rocks. Albeit some sedimentary rocks may be very hard and contain minerals similar to an igneous rock, an important decisive factor for discrimination between an igneous rock and a sedimentary rock is that the igneous rock has interlocking crystals, whereas the sedimentary rock has discreet rounded or angular grains bounded by cement.

Only a small percentage of Tennessee's exposed bedrock consists of igneous rocks, and these are actually metamorphosed igneous rocks (see figure 3.7—igneous rocks that have been metamorphosed). Exposures are located mainly in the northeastern Tennessee counties Carter, Johnson, and Unicoi.

Sedimentary rocks, because of their stratification or layering, are truly like the pages of an Earth history book. These are rocks composed of material that has been mechanically or chemically eroded from preexisting rock, mechanically transported and deposited, or deposited through some biochemical process. Depending on the origin of sediment, fossils are often associated with sedimentary rocks. Attempting to glean Earth history from a sedimentary rock becomes a forensic activity. The rock must be examined to determine the provenance (source)

		CHEMICAL COMPOSITION			
		% Silica (SiO_2) (Mostly as in quartz)			
		70%	60%	50%	40%
	Fine–grained Extrusive ("Volcanic")	Rhyolite	Andesite	Basalt	
	Coarse–grained Intrusive ("plutonic")	Granite	Diorite	Gabbro	Peridotite

Figure 3.2. A simplified scheme for classifying igneous rocks.

of the sediment, how the sediment arrived at the site of deposition, and what the conditions were at site of deposition. For example, questions posed might include: was the sediment derived from a high mountainous area, from a volcano (geosphere), or from a glacier (cryosphere); was the sediment transported by a glacier, by a slow moving meandering river (hydrosphere), by the wind (atmosphere), or by organisms (biosphere); and was the depositional site a deep ocean bottom, a beach with barrier islands and lagoons, a delta with flood plains and bayous, a desert, a swamp or marsh, a lake bottom, or a rift valley? Numerous features of sedimentary rocks can be studied in order to provide clues to answer these questions.

Mechanically formed sedimentary rocks are called **clastic**, meaning that the rocks are composed of **clasts** (broken rock fragments) that have been eroded, transported, and deposited as sediment by some agent such as wind, water, ice, or gravity. Clastic rocks or sedimentary deposits are typically classified by their textures—the sizes, shapes, and interrelationships of the particles in the rock or sediment. Grain size is used regardless of the mineral composition of the particles. Most classification schemes include the following grain sizes (from fine to coarse): clay (mud), silt, sand (which ranges from very fine to very coarse), granule, pebble, and cobble. Figure 3.3 is a simplified scheme for classifying sedimentary rocks. Although quartz, owing to its low solubility and high hardness, is the most common ingredient of sand, sand is a particle size and does not denote the type of mineral such as quartz. For instance, the sand at White Sands National Monument in New Mexico is actually the mineral gypsum ($CaSO_4$). Mud forms **shale**, silt forms **siltstone**, sand forms **sandstone**, and gravel (granules, pebbles, or cobbles) forms **conglomerate**.

Grain shapes provide clues about transportation. Roundness and angularity often hint at the distance traveled by the sediment, the transporting agent, or the energy at the deposition site, i.e., turbulence. Likewise, the sorting of grains can

			HOW FORMED		
			Mechanical (clastic)	Biochemical or Chemical (non-clastic)	
Fine		Mud	Shale	Calcium Carbonate	Limestone
↓		Silt	Siltstone	Calcium, Magnesium Carbonate	Dolomite
		Sand	Sandstone	Calcium Sulfate	Gypsum
Coarse		Gravel	Conglomerate	Sodium Chloride (salt)	Salt

Figure 3.3. A simplified scheme for classifying sedimentary rocks.

tell something about the energy in the depositional environment. Well-sorted sediment means the particles are all essentially the same size, implying that a low-energy or relatively quiet depositional environment allowed winnowing to occur. Poorly sorted sediment would likely be caused by rapid deposition without winnowing.

The process of converting sediment (formed mechanically or biochemically) into sedimentary rock is called **lithification** (also called diagenesis). It involves **burial, compaction**, and **cementation**. The particles in most sedimentary rocks are cemented together by silica, iron oxide, or the mineral calcite deposited by mineral-bearing water that permeated through the sediment.

Nonclastic rocks are those primarily formed biochemically. Direct chemical precipitation can take place in certain depositional environments, especially in an environment where evaporation is prevalent. Such deposits, called **evaporites**, include **limestone** (calcite), **dolomite, rock salt, gypsum** (calcium sulfate), and **iron oxide**.

Organisms either directly or indirectly are responsible for most of the nonclastic sedimentary rocks. Build-ups of organisms like reefs often form layers in the sedimentary record. Organisms like coral, bryozoa, or cyanobacteria are notable architects in the formation of carbonate (limestone/dolomite) strata, but the myriads of shell-bearing organisms inhabiting shallow seas are perhaps the major contributors to carbonate rock formation. The shells themselves as fossils or the mud generated from their disintegration or decomposition comprise most of the limestone and dolomite in the rock record.

The most characteristic feature of all sedimentary rocks is **stratification**, the fact that the rocks occur in layers or beds. Each bed or stratum may contain other sedimentary features that provide **paleoenvironmental** (ancient environmental) information concerning the formation of a sedimentary rock. Some common features include cross-bedding, mudcracks, ripple marks, raindrop impressions, and salt hoppers.

If sediment is temporarily exposed to the atmosphere during deposition, **desiccation** (evaporation of water) causes sediment to shrink, forming **mudcracks** that are then preserved when covered by subsequent layers of sediment (see figure 3.4). **Ripple marks** and **cross-bedding** are features resulting from the movement of the agent (water or wind) during deposition (see figure 3.5). **Salt hoppers** are the casts of salt crystals formed in an environment of super saline water. The salt crystals are subsequently dissolved, creating hollow molds that later become filled with sediment, creating cubic-shaped casts of the salt crystals (see figure 3.6).

Metamorphic rocks are preexisting rocks that have been compositionally or structurally changed by the effect of heat, pressure, chemically active gases

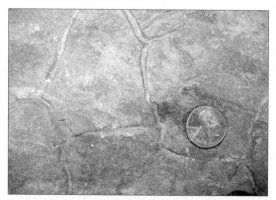

Figure 3.4. An example of typical mudcracks found in the Cambrian Rome Formation. Mudcracks form when sediment is exposed to drying conditions.

Figure 3.5. Large ripple marks in a tilted bed of sandstone in the Cambrian Chilhowee Group, near Erwin, Tennessee. Most ripple marks found in Tennessee rocks are similar in form but considerably smaller in scale.

Figure 3.6. Salt hoppers in sandstone of the Cambrian Conococheague Formation of the Knox Group, near Greeneville, Tennessee. Cubical salt crystals formed by evaporation of seawater become dissolved to form cubical molds that are later filled with sediment to form casts of the salt crystals.

or fluids, or any combination thereof. In other words, a rock that formed in an environment under one set of physical-chemical conditions will reach equilibrium when subjected to the physical-chemical conditions of a different environment. Generally speaking, however, the changes regarded as metamorphic are wrought by increasing temperatures and or pressures rather than decreasing temperatures or pressures, as in the case of weathering (a surficial process).

The primary factors determining what type of metamorphic rock forms are: 1) the **protolith** (parent or original rock being metamorphosed); 2) the levels of heat and or pressure; and 3) the presence or absence of fluids. The classification of metamorphic rocks is not as straightforward as the two other classes of rocks. Metamorphic rocks may be classified into **high** or **low grade** or **rank** based on the intensity of heat and/or pressure. Also, metamorphic rocks can be classified on the basis of texture. **Foliated** (leaflike) rocks are banded—generally exhibiting the facility to split along parallel surfaces. **Nonfoliated** rocks are massive, typically resulting from the recrystallization of the minerals in a monomineralic protolith (original rock composed of one kind of mineral).

Metamorphism can be pictured by imagining a deposit of mud being buried deeper and deeper within the Earth (increasing temperature and pressure). The mud on initial compaction becomes the sedimentary rock, shale. Then, if subjected to a temperature and/or pressure increase, the clay particles in the shale recrystallize into very fine interlocking micaceous crystals to form a dense foliated rock called **slate**. As the crystals continue to grow with increasing heat and/or pressure, though not visible to the naked eye, they give the rock a sheen in reflected light creating the foliated rock called **phyllite** (leaflike). Finally, at the high-grade level, when mineral crystals become large enough to identify with the unaided eye, the rock is called **schist**.

Gneiss is a high-grade metamorphic rock formed when minerals of the protolith partially melt and migrate into bands of alternately light- and dark-colored minerals (if totally melted, magma would be formed).

Mineralogy also can be used to ascertain the grade or rank of metamorphism. Certain metamorphic minerals form only when temperature and pressure reach a certain value. These are called **index minerals** (rock thermometers). Geologists mapping in areas that have been regionally metamorphosed, like the Blue Ridge in Tennessee, make note of the first appearance of each of these minerals and draw lines representing "equal metamorphism," which are referred to as **isograds**. The areas between isograds are referred to as **metamorphic zones** and are named for the index mineral occurring within each zone. For example, beginning with low-grade metamorphism, one would cross the chlorite isograd into the chlorite zone, then cross the biotite isograd into the biotite zone, followed in order by the garnet zone, staurolite zone, and the sillimanite zone. By recognizing metamorphic zones, geologists can infer how deeply rocks may once have been buried in the Earth, the effects of a pluton on rocks, or the proximity of rocks to the forces of rock deformation, as in plate tectonics.

Nonfoliated rocks are massive, resulting from the recrystallization of the protolith minerals. Two common nonfoliated metamorphic rocks are quartzite and marble. In the case of quartzite, the protolith is quartzose sandstone, and, in the case of marble, the parent rock is either limestone or dolomite.

Metamorphic rocks form the bedrock of the Unaka Mountains or Blue Ridge province of eastern Tennessee (see figure 1.1). The **rock cycle** illustrates the relationships existing among rock types during interactions between the geosphere and the other subsystems. Figure 3.7 illustrates the rock cycle.

Earth Processes

Perhaps the only thing constant about the Earth is that it continually undergoes change. Processes of change may be internal or external, and it is not uncom-

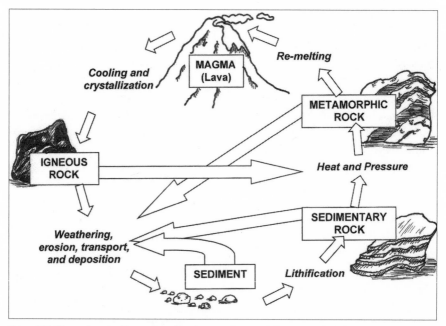

Figure 3.7. The rock cycle. The arrows show how earth materials are recycled. Seldom is the outer cycle a direct continuous sequence from magma back to magma. Instead, as the inner arrows indicate, there are "shortcuts," or smaller cycles, within the grand cycle. Note how the processes of weathering, erosion, transportation, and deposition interrupt sediment from becoming sedimentary rock, sedimentary rocks from being metamorphosed, and metamorphic rocks from being remelted into an igneous rock.

mon for the processes to facilitate or impact one another. For example, internal changes induce plate tectonics, an external activity, which in turn can generate external phenomena in the form of mountains or volcanoes.

Internal Processes

Plate tectonics, described in chapter 2, result from mantle convection currents forcing plates of lithosphere to diverge or converge. Divergence creates **rift valleys** such as the Red Sea of Africa, and convergence typically causes crustal deformation in the form of mountains (see figs. 3.8A and 3.8B). The trailing edge of a continent located on a migrating plate is referred to as a **passive margin** and where plates converge is an **active margin** (see figure 3.8C). The east coast of the North American continent is a good example of a passive margin, and the west coast is an example of an active margin. However, when two plates collide at a convergent plate margin, the conditions are not so placid. When plates collide, one plate (most often dense ocean crust) will generally slide beneath the other plate (lighter continental crust) in an action called **subduction** (see figure 3.8C). Earth material caught up in this subduction is intensely deformed into a mountainous terrain called an **orogen**. Although not all volcanoes and earthquakes are related to subduction, they are common products.

Rock deformation results when earth materials are subjected to pressures that tend to pull them apart or push them together. Features created by rock deformation range from microscopic in size to those spanning many miles. **Compression** is the force that pushes materials together, and **tension** is a force that pulls materials apart. Depending on the temperature and pressure of the environment (i.e., near the Earth's surface or buried deeply within the Earth), rocks may behave as a plastic (puttylike) at depth or brittle (like most of the hard rocks now at the Earth's surface). In a general sense, when a force is first applied to earth materials, they initially behave elastically, meaning that they change shape, and when the force is removed, they rebound to their original shapes. However, material becomes permanently deformed when the deforming force exceeds the elastic limit. The materials are deformed like taffy until the limit of plasticity is exceeded and the materials fail or break. Rocks can become permanently folded in the realm of plastic deformation and develop fractures in the form of faults, joints, or rock cleavage when the plasticity limit is exceeded. Figure 3.9 shows rocks that have been deformed plastically.

Two of the most common folds are anticlines and synclines (see figure 3.10). An **anticline** is where rock layers are arched upward. Because of superposition, after some erosion of the land surface, older layers are exposed at the center of

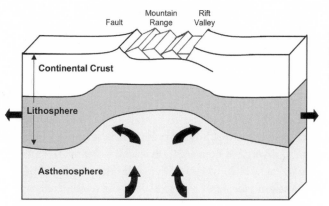

Figure 3.8A. A block diagram illustrating plate divergence (rifting) within a continent.

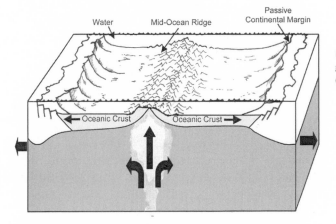

Figure 3.8B. A block diagram illustrating divergence (rifting) of oceanic plates.

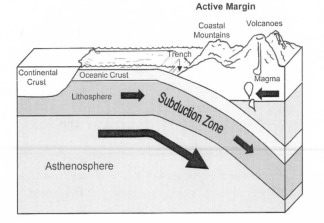

Figure 3.8C. Subduction at plate convergence.

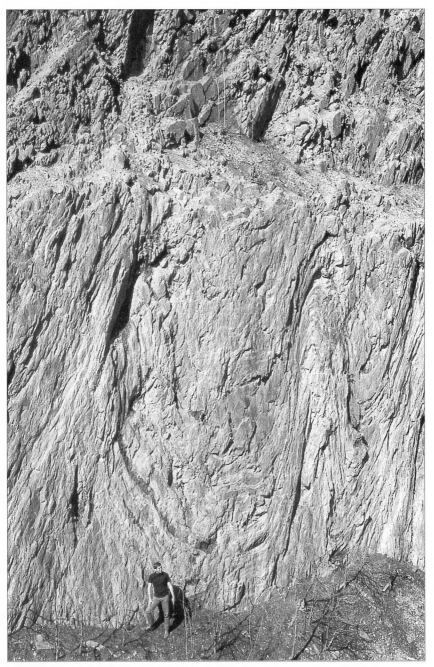

Figure 3.9. Rocks deformed plastically into folds. Note thickening in the crests and the troughs. Folds are in Precambrian Walden Creek Group rocks exposed along U.S. Highway 321, west of Townsend.

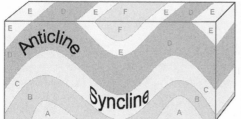

Figure 3.10. A block diagram illustrating up-arched rock layers forming an anticline and down-warped rock layers forming a syncline. Note that on the side of the block representing the land surface, the oldest layer of rock (D) forms the center of the anticline, and the youngest rock (F) forms the center of the syncline. The folds are caused by compression or squeezing.

the anticline and younger layers on the flanks. Where rock layers are bent down forming a **syncline**, the youngest layer is exposed in the core of the fold and flanked by older layers (see figure 3.10).

Faults are breaks in rocks where one rock mass slides across another; the rupturing surfaces are called **fault planes**. Most fault planes are oblique to the surface of the Earth. The mass of rock above the plane is called the **hanging wall,** and the mass below the plane is the **footwall.** When the hanging wall has moved upward relative to the footwall, a fault is caused by compressional forces, and

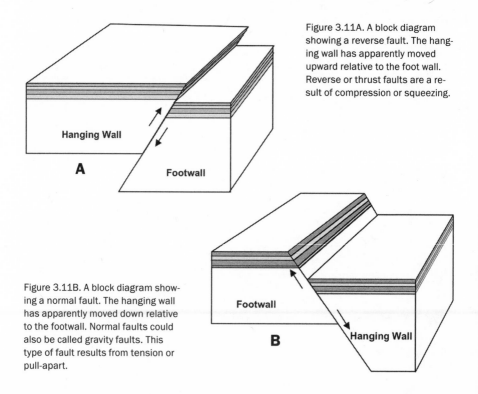

Figure 3.11A. A block diagram showing a reverse fault. The hanging wall has apparently moved upward relative to the foot wall. Reverse or thrust faults are a result of compression or squeezing.

Figure 3.11B. A block diagram showing a normal fault. The hanging wall has apparently moved down relative to the footwall. Normal faults could also be called gravity faults. This type of fault results from tension or pull-apart.

this is termed a **thrust** or **reverse fault** (see figure 3.11A). When the hanging wall has apparently moved downward relative to the footwall, it is a **normal** or **gravity fault** caused by tensional forces (see figure 3.11B). **Joints** are rock fractures with no appreciable movement along the fracture surface. They commonly are found in sedimentary rocks where they are usually perpendicular to bedding. Figure 3.12 is a recumbent fold—an anticline that has been turned on its side. Some of the many partings in the rock layers are joints, but some partings, such as those parallel with the axial plane of the fold, are called rock cleavage. These are formed as the rock layers slide across one another. Figure 3.13 shows an anticline with several sets of joints. A group of mutually parallel joints form a set, and two or more sets form a joint system. Joints play an important role during erosion (geosphere, hydrosphere, biosphere, atmosphere, and cryosphere

Figure 3.12. A recumbent fold in the Precambrian Walden Creek Group exposed along U.S. Highway 129 a short distance east of the junction with the GSMNP Foothills Parkway. The layer with the white streaks is a carbonate rock. The streaks or veins fanning outward are joints filled with quartz and calcite, and the near horizontal veins, also filled with quartz and calcite, are rock cleavage partings.

Figure 3.13. An anticline in the foothills of the Great Smoky Mountains exhibiting several sets of joints.

interactions). Many landslides such as those along I-40 between Hartford and Waterville, in Cocke County, Tennessee, are associated with the presence of joint systems. However, erosion controlled by joints created many scenic features of the landscape (see Devil's Racetrack, natural bridges, and other formations discussed in other chapters).

The rock record is punctuated by surfaces of erosion or nondeposition called **unconformities**. In the course of geologic events, bedrock or unconsolidated sediment may be exposed to erosion during a regression of the sea and then subsequently submerged during a transgression to receive more layers of sediment. The rough, usually uneven, erosional surface between the older bedrock and the younger sedimentary strata is the unconformity. Unconformities form a **hiatus** (a gap in time) in the geologic record. They mark the places in Tennessee's geologic "history book" where pages, paragraphs, or chapters are missing. It is difficult to determine how much of an older geologic record may have been removed by erosion or even whether it was ever present. It is fortunate, however, that the records missing at one location may be found in another place.

There are several varieties of unconformities. If the strata above the erosional surface are essentially parallel to the stratification below, it is called a **disconformity**. It is a **nonconformity** if the rocks below are igneous or metamorphic and the rocks above are sedimentary. It is an **angular unconformity** if the strata below, because of rock deformation, have an angular relationship with the strata above. Figs. 3.14A, B, and C illustrate the common types of unconformities.

Volcanism includes phenomena related to magma and lava and the formation of igneous rocks (see the section on igneous rocks above). Modern volcanoes and lava flows are obvious forms of volcanism; however, other than the presence of various forms of plutons, past volcanism also can be deciphered through indirect evidence. As cited in chapter 2, volcanic ejecta in the form of ash may be transported great distances, and although all vestiges of an ancient volcano may have long since been eroded away, the ash that erupted from it may be preserved in layers of the sedimentary rock record formed during the time frame of the eruption. Such is the case for the rock record in Tennessee. Layers of altered volcanic ash called **bentonite** (a mix of clay minerals) are part of the record interbedded within 400-million-year-old sedimentary rocks (see figure 10.2).

External Processes

The major external processes on Earth are weathering and gradation—the processes most responsible for what we see in the present landscape. **Weathering** is the physical and/or chemical changes in earth materials at or near the land surface. The importance of weathering in Earth processes can be noted in the rock

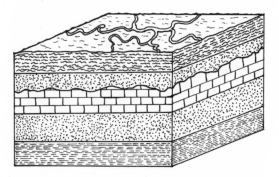

Figure 3.14A. A disconformity. The rock layers above and below the unconformity, the irregular wavy line, are essentially parallel. Only a surface of erosion or nondeposition separates the upper and lower layers.

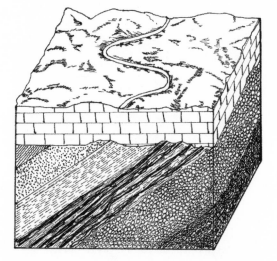

Figure 3.14B. An angular unconformity. The rock layers above and below the unconformity, the irregular wavy line, form an oblique angle indicating that the rock layers below were deformed and eroded prior to the deposition of the upper layers.

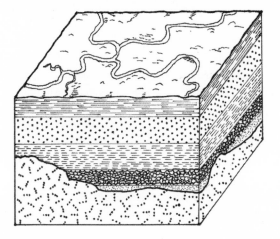

Figure 3.14C. A nonconformity. The rocks below the unconformity are igneous or metamorphic and were most likely eroded prior to the deposition of the sedimentary layers above the unconformity represented by the irregular wavy line.

Figure 3.15. Spheroidal weathering of Precambrian Cranberry Granite Gneiss near Hampton, Tennessee. The more rapid weathering along a system of joints causes layers of rock to spall off like layers of an onion.

cycle (see figure 3.7). **Gradation** is either **degradation** (erosion) or **aggradation** (deposition) by the geologic agents wind, water, ice, gravity, or organisms. These processes have sculpted the Earth's surface for eons and continue to shape the environment today.

Physical/mechanical weathering is disintegration, simply the breaking of materials into smaller and smaller pieces by physical forces. Water freezes when it expands, and if it is within rock partings, it can readily split the rock. Also, tree roots embedded in rock crevices split rock as the tree grows larger and larger.

Chemical weathering is the decomposition of earth materials under conditions of low temperature and pressure (as compared to metamorphism, which alters rocks at high temperatures and pressures). The rust that forms on metal objects left outdoors is an example of a chemical weathering process called oxidation. One of the most effective agents of chemical weathering is a ubiquitous, mild acid called carbonic acid, formed in the atmosphere through the combination of water and carbon dioxide. Though mild, it reacts with minerals to create new minerals. Most clay is the product of carbonic acid decomposing various silicate minerals—especially feldspars and micas. Figure 3.15 is a type of chemical weathering called **spheroidal weathering** where layers of the rock are spalled off in layers like an onion.

Figure 3.16. Chemical weathering of Cranberry Granite Gneiss has produced a deep regolith of mineral grains destined to become grains in a sedimentary rock. Note the rock hammer sunk into the regolith. Photograph taken a few feet from fig. 3.15.

Figure 3.17. Knox residuum with an armor of chert debris. The soil is typically reddish-orange or yellowish orange, and the chert, which is referred to as "float," is generally buff in color but may have bands of gray or black. The concentric banding is interpreted to represent silicified remains of cyanobacteria that formed stromatolite reef structures that served as frameworks for deposition of many carbonate rocks in the Knox Group. Other chert is chemically deposited silica that in some cases has been leached from volcanic ash deposits (see chapter 10).

Loose, unconsolidated material at the Earth's surface is called **regolith** (see figure 3.16). Regolith through the effects of climate, topography, organic matter, and time forms soil. Chemical weathering, especially in warm, humid climates, produces a regolith known as **residuum** consisting of insoluble residues formed in situ by leaching. Reddish soils containing varied colored fragments found across much of Tennessee were formed from residuum weathered from **chert**-bearing (a form of silica) carbonate rocks (limestone or dolomite). Figure 3.17 is typical chert-bearing residuum from the Knox Group.

Some regolith has been transported to its location. For example, regolith formed from landslide debris is called **colluvium**, and sediment transported by streams and rivers is called **alluvium**. Colluvium is an unsorted deposit of rock fragments, typically angular in shape, that may range in size from clay up to boulders. Alluvium, on the other hand, having been transported by water, is generally well sorted and finer-grained.

An understanding of present-day external processes of change provides us with analogs that can be used to interpret the records of Earth history or as James Hutton put it—"the present is the key to the past."

4

Geologic Time

> As there is not in human observation proper means for
> measuring the waste of land upon the globe, it is hence in-
> ferred, that we cannot estimate the duration of what we see
> at present, nor calculate the period at which it had begun;
> so that, with respect to human observation, this world has
> neither a beginning nor an end.
>
> —James Hutton, Scottish geologist,
> *Abstract of a Dissertation* (1785)

Although the geologic history presented in this book treats only the last billion years of Earth history as recorded in Tennessee rocks, the Earth is at least 4.5 billion years old. This begs the question—how do we know? First, consider that when James Hutton, author of the principle of uniformitarianism, recognized that the sedimentary rock record was created by processes like those occurring in the present, he could surmise only that the Earth had a long history. During his moment in human history, the Earth was believed to be no more than 4000 to 6000 years old, hardly sufficient time to complete the events he envisioned at Siccar Point, Scotland. Possibly what might have gone through Hutton's mind, as he pondered the rocks he saw, was that there had to be enough time to accumu-late a thick sequence of rock, deform the rocks into mountains, erode the moun-tains, and then deposit another thick sequence of rocks upon the eroded surface of the mountains. Means of reckoning geologic time have evolved through the years with improved technology and especially the knowledge of radioactivity. If Hutton had only known what we know today, he would have been a very happy man. Two basic ways are used to assign the timing of events in the geologic re-cord. The most basic scheme for timing past geologic events is actually very sim-ple; it is called **relative time**, which merely places events into a chronological sequence without reference to any specific time in the past. The other method is **absolute time,** which places a specific date on an event.

Relative Time

This method could be referred to as the **principle or rule of sequence**. Geologic events are sequenced as to whether they occurred *before* or *after* a particular reference. It is sort of like the relative expressions yesterday and tomorrow, which are terms referenced to today. A basic principle of relative dating is known as **superposition**. This principle states that, in a sequence of sedimentary layers, the lowermost layer is the oldest, and each layer becomes progressively younger toward the top. A corollary to superposition, used to determine timing of rock deformation, states that the age of deformation is younger than the age of the youngest rock deformed and is possibly older than the oldest rock not deformed. Likewise in the case of dating igneous intrusions, the age of an intrusion is younger than the youngest rock intruded or affected (metamorphosed) and is possibly older than the oldest rock not intruded or affected. This is sometimes referred to as the principle of cross-cutting relationships. Fossils too can be used for relative dating. Species of organisms generally live for only a few million years before becoming extinct; thus, if it is known that a certain species lived before or after other species, it can be determined that sedimentary layers are older or younger than other layers.

Absolute Time

All matter consists of atoms, and an atom is made of a nucleus of particles called protons and neutrons that is surrounded by electrons. Every known atom (element) is identified by the number of protons in its nucleus. A normal atom has the same number of electrons (negatively charged particles) as it has protons (positively charged particles), which provides electrical balance. Gaining or losing electrons can cause an atom to take a negative or positive charge, respectively. However, gaining or losing neutrons (no electrical charge) from an atom affects only an atom's mass or weight with no affect on its electrical balance. Elements (atoms) with different numbers of neutrons are called **isotopes**. The element carbon, for example, has three isotopes, each with a different mass or weight—carbon-12, carbon-13, and carbon-14. Most common chemical elements have several isotopes; however, certain isotopes tend to be unstable and are considered to be **radioactive**. This means that regardless of temperature or pressure, these unstable or radioactive isotopes spontaneously disintegrate (lose mass or weight) to form more stable isotopes. The rate of disintegration is measured in half-lives. One **half-life** is the amount of time required for half of a radioactive isotope to decay to another stable isotope—e.g., uranium-235 decays to lead-207 with a half-life of about 700 million years. Measuring geologic time is based on the ratio of the amount of parent isotope (e.g., uranium-235) to the amount of decay product or daughter isotope (e.g., lead-207). For example, after 700 million years, a half-gram would remain of a one-gram sample of uranium-235, and one-fourth of a gram would remain after 1400 million years. Most isotopes used to date rocks within the geologic record have long half-lives and are useful for dating events millions to billions of years in the past. However, an organic isotope, carbon-14, is used in dating more recent geologic events, as in archaeological investigations where materials are less than seventy thousand years old. The half-life of carbon-14 is about 5730 years. In radiocarbon dating, the amount of parent material left is measured and compared to a universal level of carbon-14 in the present atmosphere rather than the amount of a daughter product formed, because carbon-14 decays to an isotope of nitrogen which is not retained. So, after 5730 years, one-half of the level remains, after 11,460 years, one-fourth of the level remains, until after 70,000 years there is an insufficient amount of carbon-14 left to measure. Figure 4.1 illustrates several examples of relative dating, as well as how absolute dating can be used to establish a chronology of geologic events. The geochronologic methods presented here are not the only available methods used in dating past events, but they are good examples of how geologists go about establishing a chronology of Earth history.

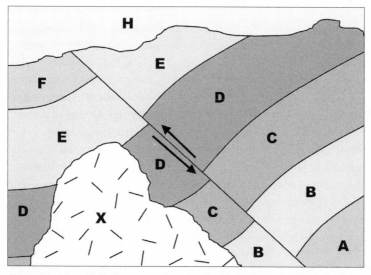

Figure 4.1. A hypothetical cross-section of a slice of the Earth showing a number of geologic events. Superposition shows that Bed A is the oldest and Bed H is the youngest. Cross-cutting relationships indicate that the fault (fracture with arrows showing relative movement) is younger than Bed F and older than Bed H; and the igneous intrusion (X) is younger than Bed E and possibly older than Bed F. If the radiometric date of the igneous intrusion was determined to be 200 Ma, then Beds A–E are older than 200 Ma, and the fault is younger than 200 Ma.

Geologic Time Scale

The division of geologic history into units is a story unto itself. Suffice it to say, there has been considerable evolution of thought involved. Well before radiometric dating was established, units were primarily based on the dominant forms of life that occurred at various times in the record. These units are still recognized today, even though they do not relate directly to units of absolute time. The broadest units of the record are called **eons**. At present, four eons are recognized. The Hadean and Archean Eons span almost half of geologic time (from at least 4.5 billion years ago to 2.5 billion years ago) and include fossil primitive bacteria and algae. The Proterozoic Eon (from 2.5 billion years ago to 542 million years ago) contains fossils of some of the earliest aquatic plants and covers the earliest part of the Earth's story. The Phanerozoic Eon covers from 542 million years ago to the present and marks the time when complex forms of life rather abruptly evolved and flourished.

Eons are further subdivided into **eras**, which in turn are further divided into **periods**. Because the first period of the Phanerozoic Eon is the **Cambrian Period**, all older rocks are often referred to as **Precambrian** (belonging to the Hadean, Archean, and Proterozoic Eons). From oldest to youngest, the eras of the Phanerozoic are Paleozoic, meaning "ancient life," Mesozoic, meaning "middle life," and Cenozoic, meaning "recent life." In general, marine invertebrate organisms dominated during the Paleozoic; dinosaurs and trees were dominant in the Mesozoic; and mammals were and are prevalent in the Cenozoic.

The periods of the time scale, especially in the Paleozoic Era, are typically named for a geographic area in the world where the rocks representing that period were first studied in detail or some other cultural aspect of history. As examples, for the Paleozoic Era, the Cambrian Period is named for an area in Wales; the Ordovician and Silurian Periods come from the names of early European tribes; the Devonian from Devonshire; the Mississippian from the upper Mississippi River valley; the Pennsylvanian from the state of Pennsylvania; and the Permian from Perm, a Russian region near the Ural Mountains. Figure 4.2 is a generalized geologic time scale.

Many schemes have been devised to help visualize the immensity of geologic time in relation to human development. One method is to visualize 4.5 billion years fitting onto a football field. In returning a kickoff from one goal line, we would pass midfield and cross our opponent's 12-yard line at the end of Neoproterozoic time (542 million years ago); observe the extinction of dinosaurs (65 million years ago) as we cross between the one- and two-yard lines; four inches from the goal line we would see our first human ancestors (about 5 million years ago); and five thousandths of an inch (0.005 inch) from crossing the goal, not even the thickness of the "pimples" on the football, we could study the beginning of recorded history (about 6000 years ago).

Eon	Era	Period	Epoch	Age Ma
Phanerozoic	Cenozoic	Quaternary	Holocene	0.01
			Pleistocene	2.6
		Tertiary — Neogene	Pliocene	5.3
			Miocene	23
		Tertiary — Paleogene	Oligocene	33.9
			Eocene	55.8
			Paleocene	65.5
	Mesozoic	Cretaceous		145.5
		Jurassic		201.6
		Triassic		251
	Paleozoic	Permian		299
		Carboniferous	Pennsylvanian	318
			Mississippian	359
		Devonian		416
		Silurian		444
		Ordovician		488
		Cambrian		542
Precambrian	Proterozoic	Neoproterozoic		1000
		Mesoproterozoic		1600
		Paleoproterozoic		2500
	Archean			3850
	Hadean			?

Figure 4.2. Geologic Time Scale. Ma = Millions of years before the present (modified from Walker, Geissman 2009)

5

Giants of Tennessee Geology

The important thing in science is not so much to obtain
new facts as to discover new ways of thinking about them.
—Sir William Bragg, British physicist (1862–1942)

As stated earlier, one of the variables in the equation of geologic history in Tennessee is the perceptiveness of the observers of the state's bedrock. Deciphering the geologic history cannot be attributed to the perceptions of any one geologist. Many workers have been involved in making the countless geologic observations necessary for deciphering and compiling the geologic history of Tennessee. The goal of those initial investigations was not to determine geologic history, but to seek natural resources for a growing country. Mineral exploration often involved working out the sequence of geologic events; hence a geologic history evolved.

Interestingly enough, the rock exposures scrutinized by the earliest Tennessee geologists have not changed drastically since they made their observations. However, the interpretations of Tennessee geology have continuously changed through time. Advantages each new generation of geologists have had over their predecessors include more outcrops to measure and interpret (e.g., new highway exposures), new technologies—e.g., boreholes, Global Positioning System (GPS), and Geographic Information Systems (GIS), and a variety of other geophysical implements, better base maps, the advancement of new ideas and theories in geology, such as plate tectonics, and certainly motorized vehicles, which have an advantage over horseback.

The bottom line is simply that this book is not solely based on the writer's observations and interpretations, but on that of legions of workers—past and present. Hence this text is built on the labor of many giants of geology.

The era of "modern geology" began overseas with giants such as Scotland's James Hutton (1726–1797), considered, as previously mentioned, the founder of "modern geology"; England's Charles Lyell (1797–1875) and William Smith (1760–1839); and Louis Agassiz (1807–1873), who was born in Switzerland. Also among this coterie of early international naturalists was Tennessee's first state geologist—Gerard Troost (1776–1850). Troost, born in the Netherlands, arrived in the United States in 1810. One of the moves he made prior to entering Tennessee's geologic scene was to join a communal experiment in New Harmony, Indiana. William Maclure (1763–1840), a Scotsman like James Hutton and considered a geologist of considerable fame, enticed Troost in 1826 to join with another pioneer American geologist, David Dale Owen, to develop a community dedicated to science at New Harmony. Troost was hired at $500 per year to teach mathematics and science. Although the concept of New Harmony was noble, for various reasons the social experiment failed in 1827.

An acquaintance from Troost's past, in Philadelphia, who was by then teaching chemistry at the University of Nashville, enticed Troost to join him. Through his diligent fieldwork, teaching, and research, Troost became Tennessee's first state geologist in 1831. As a teacher, Troost was a strong advocate of hands-on field studies, and the format of his sequence of physical and historical geology courses soon spread to other colleges and universities. Many state geologists were to follow Gerard Troost, but James Safford (1822–1907) was perhaps the most notable (Corgan 1985; Corgan and Gibson 1991; and Corgan and Gibson 1996).

Although Safford is best known for geologic investigations of Tennessee, he also taught chemistry and biology, as well as geology, for fifty-two years. Safford's administrative posts were primarily in medicine, as was the case for many nineteenth-century earth scientists, and even his teaching of geology was mostly to would-be physicians, engineers, and pharmacists. His deep-seated interest in geology was no doubt cultivated in Silliman's laboratory at Yale in 1847. Benjamin Silliman (1779–1864) was the first science professor at Yale and the founder of the *American Journal of Science.*

From 1848 to 1873, Safford held various teaching positions at Cumberland University in Lebanon, Tennessee; however, he took leave from that tenure in 1854 to perform a geologic survey of Tennessee while serving as state geologist. Following the Civil War interlude, Safford held several part-time teaching jobs, including at Vanderbilt University, continued to maintain his farm, investigated oil and gas properties, and in 1869 published his seminal work, *The Geology of Tennessee.*

While serving a half-time appointment as professor of geology and biology at Vanderbilt University, Safford taught chemistry in a medical department jointly operated by Vanderbilt and the University of Nashville. Also during this time, Safford collaborated with J. B. Killebrew, another Tennessee scientist who later became state geologist, to publish a popular geology textbook, *Elementary Geology of Tennessee,* which went through new editions at least through 1900. A competing textbook by W. G. McAdoo and H. C. White bearing the same title, *Elementary Geology of Tennessee,* was published in 1875. These two books, along with other earth science-related books, brought about somewhat of a renaissance to the subject of geology in Tennessee colleges and secondary schools.

Since the Tennessee Geologic Survey was officially established in 1909, many state geologists have followed in the footsteps of Troost and Safford, and along with many other Tennessee Division of Geology geologists, they have contributed to a better understanding of Tennessee geology. Likewise, many geologists from academia, the United States Geological Survey (USGS), and the Tennessee Valley Authority (TVA) scrambling over the hills and vales of Tennessee have made significant contributions. Although these geologists have basically observed the same rocks over the years, each new generation of geologists has been able to make their observations through glasses tinted by enhanced technology and the evolution of thought concerning Earth processes, i.e., plate tectonics. Not only have base maps for plotting geology improved over time, but now there are surveying techniques that utilize various types of remote sensing as well as GIS.

The next group of geologists who made tremendous contributions to understanding Tennessee were members of the USGS field parties engaged in producing the classic geologic folio series. A folio consisted of several maps of a region portraying topography, bedrock geology, structural geology, and economic geology along with discussion of stratigraphy, agriculture, and ethnology. Three folio mappers associated with the geology of Tennessee stand out. First, C. W. Hayes published eight folios between 1891 and 1909, including Ringold, TN/GA; Kingston, TN; Chattanooga, TN; Sewanee, TN, Cleveland, TN; Pikeville, TN; McMinnville, TN; and Columbia, TN. Second, two other folios were published during that same time by M. R. Campbell, including Standingstone, TN, and Bristol, TN/VA. The third member of the team was Arthur Keith. Between 1895 and 1927 Keith published the following eight folios: Knoxville, TN; Loudon, TN; Morristown, TN; Briceville, TN; Wartburg, TN; Maynardville, TN; Cranberry, NC/TN; Greeneville, TN; and Roan Mountain, NC/TN.

Although many geologists now contribute to our understanding of Tennessee geology, a list of giants would be remiss if several more names of geologists were not mentioned. In consideration of the three grand divisions of Tennessee, the works of several geologists must be mentioned. For East Tennessee, P. B. King's USGS Professional Papers on the geology of the Great Smoky Mountains (1958–1968) and John Rodgers's Tennessee Division of Geology Bulletin 58 (1953), *Geologic Map of East Tennessee with Explanatory Text,* are most notable. The primary references for the geology of Middle Tennessee are the works of Charles W. Wilson Jr., especially Tennessee Division of Geology Bulletin 56 (1949), *Pre-Chattanooga Stratigraphy in Central Tennessee.*

One might say that West Tennessee is still a frontier as far as thorough geologic understanding is concerned. No apparent hero of West Tennessee geology has emerged. However, though not widely published, the maps produced by Ernest Russell and the work by Russell and Parks (1975) might be considered significant contributions towards that understanding. George I. Whitlach also should be given due consideration as a noteworthy contributor for his work with the Tennessee Division of Geology in the 1930s highlighting the ball clay deposits of West Tennessee (a leading natural resource of the region).

Often overlooked are the contributions to engineering and environmental geology made by geoscientists of the Tennessee Department of Transportation (TDOT). The giant in this category is David Royster, who introduced the department to the importance of geology when constructing and maintaining roads across Tennessee. He started with TDOT in 1958 as an assistant soils scientist, but his geologic background soon moved him to the head of the geotechnical section. He cut his teeth on devising solutions for the many landslides that beset the construction of I-40 up the escarpment of the Cumberland Plateau above Rockwood, Tennessee. From scratch, Royster created and developed a section with a group of expert geoscientists capable of understanding and dealing with landslides, slope stability, and **karst** (e.g., sinkholes).

6

Supercontinents, Continents, and Terranes

Simplicity is the ultimate sophistication.

—Leonardo da Vinci (1452–1519)

Before proceeding to the discussion of deep geologic time and the formation of Tennessee's rock record, the main protagonists introduced in chapter 2, supercontinents, continents, and terranes, need to be discussed in a bit more detail. As mentioned earlier, the evolution of thought leading to the plate tectonics theory, which includes the related supercontinents, continents, and terranes, has revolutionized the manner in which Tennessee's geologic history is interpreted.

Most of the facts of the plate tectonics theory had been known for nearly a century before it evolved from an hypothesis to a well-substantiated and supported theory. As described in chapter 2, plate tectonics is regarded as an important unifying theory in the geologic sciences that helps us understand how the geosphere works. Geologic research is continually improving this understanding. Still, one of the most nebulous parts of the theory is the initial stages in development of the earliest continental plates; in other words, how it all began. Despite this shortfall, chapter 2 discusses how plates on the Earth's surface have continually moved apart and collided throughout geologic time. Now, let us examine more closely the characteristics of the pieces of the Earth called plates.

Continents are the present-day landmasses that are the above-sea-level portions of plates (such as North America, South America, Africa, Eurasia) as opposed to plates (Pacific, Nazca, and Cocos) that consist predominantly of seafloor (see figure 2.3). In the course of Earth history, there were times when continents were dispersed in a fashion similar to what we see today, and then there were times when continents coalesced to form **supercontinents**. In what appears to have been almost a cyclic fashion, supercontinents have had tendencies to break up into continents that drift apart. Among the continents, or between the cracks, so to speak, smaller chunks of "real estate"—small displaced blocks of continental crust ("microcontinents" such as present-day Madagascar) or volcanic islands or submerged seamounts—move about while riding upon plates made up of seafloor. These bits of "real estate" are called **terranes**. Typically, during subduction when plates collide, terranes caught in the collision zone, rather than being subducted, are scraped off and become sutured (attached) to or thrust onto a continent (**accretion**). As mentioned in chapter 2, when the story of Tennessee history began, just about three-fourths of the Earth's history had already taken place, and plates were already moving about on the face of the Earth.

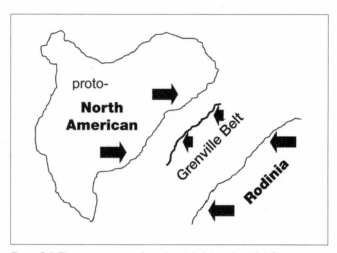

Figure 6.1. The convergence of continental plates about 1.1 Ga to create the supercontinent Rodinia and the Grenville Mountain belt.

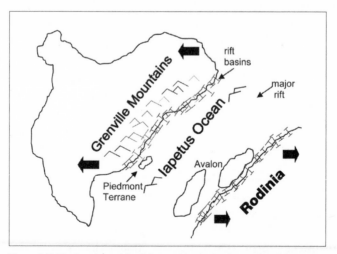

Figure 6.2. The breakup of Rodinia resulting in the formation of the vast Iapetus Ocean. Note that remnants of the Grenville Mountains were left behind on the margin of proto-North America. These remnants are the oldest rocks in Tennessee.

Our Tennessee story begins approximately 1100 Ma, when old plates containing continental masses, one of which was proto–North America, collided to form the supercontinent Rodinia (see figure 6.1). The region of the collision was intensely folded and faulted, thickening the Earth's crust to form an elongate mountainous region called an **orogen**. Such an episode of mountain building

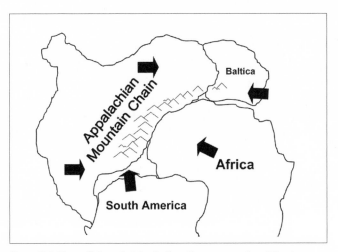

Figure 6.3. The collision of plates to form the supercontinent Pangea and the structure underlying Appalachian Mountains. The mountains formed at this time likely had the stature of the present-day Himalayan Mountains.

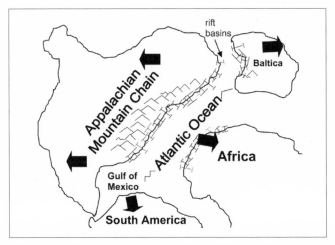

Figure 6.4. The breakup of Pangea. The divergence of plates resulted in the formation of the Atlantic Ocean.

is known as an **orogeny**. It will become evident in subsequent chapters of this book that the rock record of Earth history has been punctuated many times by orogenies (mountain-building events).

The oldest rocks in Tennessee are igneous and metamorphic rocks related to the Grenville orogeny, when Rodinia was formed by the collision. Except in the northeastern part of Tennessee, where they are exposed at the surface by erosion, these rocks are buried beneath younger rocks all across Tennessee and are frequently referred to as the "basement." Rodinia eventually split apart to form a vast ocean called Iapetus (see figure 6.2). But, in the course of history, several hundred million years later, another series of collisions with related orogenies formed the supercontinent Pangea, which in turn also eventually broke up (see figs. 6.3 and 6.4). Throughout the course of deep geologic time, the rifting, drifting, shifting, colliding, and separating of what has ultimately become our continent, North America, caused the real estate of what is now Tennessee to be subjected to many diverse environmental conditions relative to sea level and proximity to the equator. In other words, over time, the tectonics of the geosphere had profound effects upon the hydrosphere, atmosphere, biosphere, and cryosphere. These effects are preserved in the rock record as part of the geosphere and will be discussed in subsequent chapters.

7

The Record

> The past history of our globe must be explained by what
> can be seen to be happening now. No powers are to be
> employed that are not natural to the globe, no action to be
> admitted except those of which we know the principle.
>
> —James Hutton, Scottish geologist (1726–1797)

The earliest attempts to develop a geologic history used the principle of superposition (see chapter 4) to establish a chronology of events. Terms such as *primitive, transitional, secondary,* and *tertiary* were used for successive bodies of rock, from the bottom to the top, respectively. Thus time was affixed to physical units of rock, even though at that time in history the Earth was believed to be only around 6000 years old. As the science of historical geology (sometimes equated with **stratigraphy**—the branch of geology that studies layered rocks) evolved, it became obvious that units or packages of rock do not exactly correspond to time. The names of the time units discussed in chapter 4 gradually developed as more of the geologic record became known, and they are still universally used. However, the rock layers recognized by geologists mapping in the field did not always fit perfectly into the brackets of the time units, plus the fact that a package of rock representing an era or even a period of time would be immense. So, along with time nomenclature, rock unit terms evolved and eventually time-rock terms were coined.

Rock (Stratigraphic) Units

Rock units are definable bodies of rock used primarily by geologists who are compiling the land surface distribution of geologic data while mapping a region. The basic rock unit is called a **formation**. By definition, a formation is a mappable body of rock (at a map scale of 1:24000 or larger) that has well-defined upper and lower boundaries with contiguous rock units; is genetically related (all of the material in the unit appears to have the same set of environmental aspects); and is binomial (has two parts to its name). One part of the name is taken from the place where the unit is well exposed and was first described (known as the **type locality**), and the second part of the name is related to the type of rock, e.g., Chattanooga Shale, Maynardville Limestone, or Porters Creek Clay. In some cases, one or more genetically related rock types form a unit and the term "formation" is used in place of a rock name, e.g., Rome Formation, Coon Creek Formation, or Hermitage Formation. Related formations are sometimes lumped into larger units called groups and groups into supergroups. An example of a group would be the Knox Group in East Tennessee, which consists of five formations; and an example of a supergroup would be the Ocoee Supergroup in the Great Smoky Mountains; which consists of three groups.

Time-Rock Units

Time-rock units are packages of rock that, regardless of rock unit boundaries (such as formations), directly correspond to time units. So time units and time-rock units are directly related, but rock units are not at all directly related to either. What is observed in the outcrops along Tennessee road-cuts and other exposures are the rock units. Figure 7.1 shows the relationship of the three ways of referring to the geologic record. Although the time units, periods, and epochs have corresponding time-rock terms, it is not customary to refer to a time-rock unit that would correspond to an era—the package of rock would be immense. To look at these relationships another way, suppose we started shoveling sediment into a large hole first thing in the morning, filling half of the hole with sand and then covering the sand with cobbles, and finishing at the end of the day. The

TIME UNITS*	TIME-ROCK UNITS	ROCK UNITS
Era		Supergroup
Period =	System	Group
Epoch =	Series	Formation
Age =	Stage	Member

*See chapter 4 for discussion of time units

Figure 7.1. The relationships of time units, time-rock units, and rock units. See, for example, that the rocks formed within a period constitute a system and rocks formed during an epoch constitute a series, etc.

sand and the cobbles would be separated by a well-defined contact. The sand layer would equate to the "rock unit" formation, as would also the layer of cobbles. However, the two "rock units" were deposited during the "time unit" of one day; thus the sand and the cobbles represent discreet rock units, but together they form a "time-rock unit."

Trying to reconcile the physical rock record with geologic time is not a simple matter. Consider this scenario: the case of a transgressing sea where the sand accumulating on the beach becomes progressively younger as the sea transgresses or laps farther onto a continent. Trying to match time-rock units with rock units would not work too well in this instance. The sandstone formation resulting from the transgressing beach is actually time-transgressing (see figure 7.2). The sand of the beach would likely grade into mud (shale) and limestone in progressively deeper water offshore from the sandy beach, so that at any point in time, some of the sandstone is coeval with some shale and limestone. In other words, they were all three being deposited at the same time. These **penecontemporaneous** (being deposited at the same time) lateral transitions in rock types related to changes in depositional environments are called **facies**. Figure 7.3 is an actual example of facies illustrated by the Cambrian Conasauga Group.

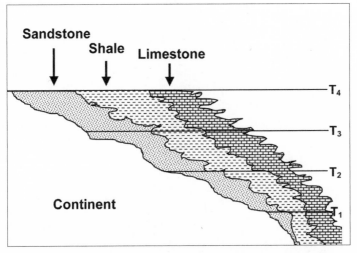

Figure 7.2. A typical sediment relationship developed in a sea transgressing onto a landmass—coarse sediment (sand-sandstone) nearest to shore, finer sediment (mud-shale) deposited farther offshore, and limestone deposited in the clearer, sediment-free water farthest offshore. T-1, T-2, T-3, and T-4 are time lines representing the transgressing sea onto a continent. Note that sediment forming the sandstone, shale, and limestone are the same age along the time lines—they are laterally equivalent and therefore facies.

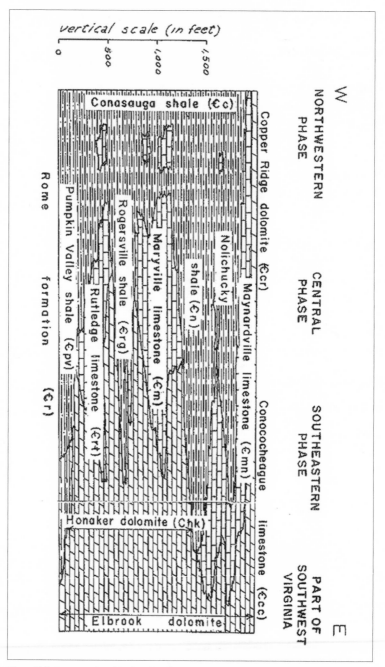

Figure 7.3. An example of facies. The Conasauga Group of East Tennessee. Note that the rock units (facies) grade from clastic on the west to nonclastic (carbonate rocks) to the east, suggesting a western source for the sediments. (Byerly 1966.)

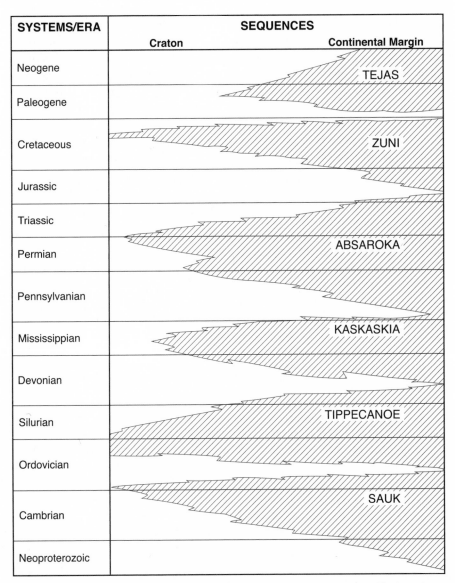

Figure 7.4. Time relationships between stratigraphic time-rock units and sequences. The jagged lines represent transgressions and regressions; the crosshatched areas are the times represented by the sedimentary rock record. Note that the time gaps lengthen toward the craton or the center of the continent. (Sloss 1963.)

Transgressions and regressions of a sea across continental areas are related to sea-level changes and, as noted above, are considered time transgressing. Because rock units almost universally tend to be time transgressing, a more reasonable method of packaging rock units based on major transgressing-regressing events has been devised. For example, the base of a rock unit package would be marked by a time-transgressing rock unit and the top boundary would be marked by the regressing rock unit or by an unconformity or erosional surface (see chapter 3). Such packages of rock have come to be known as **sequences**. L. L. Sloss first proposed the term in 1963, and although the use of the term has been modified since that time, the Sloss nomenclature is a convenient way to examine the rock record of Tennessee. Figure 7.4 shows the relationship of time or time-rock units with stratigraphic sequences. Note that the Sauk sequence ranges in geologic age from late Neoproterozoic through early Ordovician time. That means the sea began transgressing during the Neoproterozoic and regressed during early Ordovician time, leaving a break (an unconformity or erosional surface) between the early Ordovician and middle Ordovician rock record.

8

Deepest Time in Tennessee

> It is the little causes, long continued, which are considered
> as bringing about the greatest changes of the earth.
>
> —James Hutton, Scottish geologist, *Illustrations of the*
> *Earth* (1795)

More than a billion years ago, several continental plates, including what is now North America, began a series of collisions called the Grenville orogeny resulting in formation of the supercontinent Rodinia. In Tennessee the only exposures of rocks recording this event are located in the northeastern corner of the state. However, deep drilling across Tennessee would reveal their presence as the bedrock basement beneath a veneer of younger sedimentary rocks ranging in age from the Neoproterozoic Era to Cenozoic. Figure 8.1 shows the location of the bulk of the Grenvillian rocks in Tennessee in the vicinity of Roan Mountain. All these old rocks have experienced several episodes of rock deformation over time and thus bear the marks of more than one metamorphic event—a fact that makes deciphering this very early geologic history a real challenge.

Possibly the oldest Proterozoic rock in Tennessee related to the Grenville orogeny is the Cloudland or Carvers Gap Gneiss mapped as Roan Gneiss on the *Geologic Map of Tennessee* (Hardeman 1966). Using samples from the Carver Gap area near the Tennessee–North Carolina border, Monrad and Gulley (1983) dated the metamorphic event that formed the Cloudland Gneiss to be around 1.8 Ga (giga annum = billions of years (Paleoproterozoic). The rock is classed as **paragneiss,** which means that the original rock was sedimentary. This implies that an even older rock had to exist as the source for the sediment of the sedimentary rock that became the paragneiss. There is speculation that the sediment may have been derived from one of the continental plates involved in the formation of supercontinent Rodinia. Predominant minerals forming the gneissic banding are amphibole and garnet, but there are zones of mica schist and granitic minerals (quartz and feldspar). Dikes and pod-shaped sills of Bakersville Gabbro intrude the Cloudland, and as shown by Hardeman (1966), the Beech Granite Gneiss has a cross-cutting relationship with the Cloudland as well. This relationship indicates that the Bakersville and the Beech are younger than the Cloudland (see chapter 4).

Just as in the case of the Cloudland Gneiss, the origin of the next oldest of Tennessee's Grenvillian rock units, the Cranberry Granite Gneiss, is also an enigma. One hypothesis is that it originated as a granitic **batholith** (a large intrusion of igneous rock or **pluton**), and another is that it formed as a thick sedimentary deposit (which suggests, again, that there was an older rock that was eroded

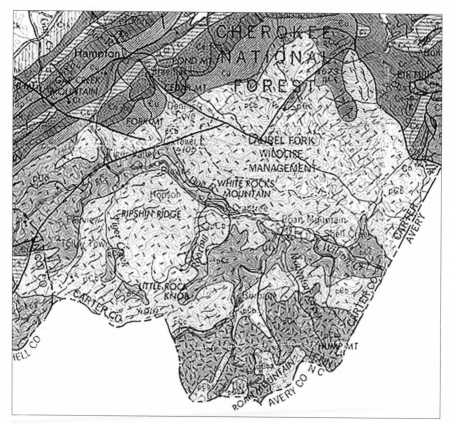

Figure 8.1. Map showing Precambrian rocks in the vicinity of Roan Mountain, Tennessee: pЄr = Roan Gneiss (Cloudland Gneiss); pЄc = Cranberry Granite Gneiss; pЄb = Beech Granite Gneiss; pЄba = Bakersville Gabbro; heavy black lines are faults; rock units striking NE-SW in the northwestern corner of the map are Paleozoic in age. (Map from Hardeman 1966.)

to supply the sediment). In either case, during the collision of plates, the original rock was intensely metamorphosed (short of being melted or remelted into magma). Although mainly composed of feldspar and quartz, the rock contains minor amounts of zircon and magnetite, as well as dark ferromagnesian minerals such as amphibole. In appearance the rock ranges from the classic banding of light and dark minerals characteristic of gneiss to an equigranular texture more typical of granite. Figure 8.2 shows a phase of the Cranberry classified as **unakite** (named for the Unaka Mountains of Tennessee). Unakite is an attractive rock characterized by blotches of pink feldspar and a pistachio green mineral, epidote. Radiometric dating by various studies place the age of the Cranberry between 1.0 and 1.3 Ga (Fullager and Bartholomew 1983). Thus the Cranberry is late Mesoproterozoic or early Neoproterozoic in age.

A problem with all the dates proposed for these rocks is that they can either reflect the time of intrusion, if it is a pluton, or indicate recrystallization of minerals during metamorphism. A dilemma arises when a pluton becomes metamorphosed. The radiometric clocks began ticking with the crystallization of the minerals containing the parent isotopes. Regardless, these rocks are very old, no matter which mode of origin is considered. If it is a pluton, there had to have been

Figure 8.2. A variety of granite called unakite. The light-colored rectangular-shaped crystals are phenocrysts of pink feldspar.

an older existing rock to be intruded (possibly the Cloudland Gneiss), and if it is metamorphosed sedimentary rock, the sediment had to have been eroded from a preexisting rock. One argument for a sedimentary origin lies in the shape or form of the mineral zircon found in the Cranberry. Zircon, which is very resistant to weathering and can survive many cycles of erosion and transportation, is often found associated with quartz in beach sediment and would be somewhat rounded from abrasion during weathering and erosion. On the other hand, if zircon crystallizes from a magma along with other minerals to form granite, the zircon will exhibit a perfect crystal form. Also, because zircon is radioactive, the crystalline zircon can be used to determine the absolute age of a pluton. However, the zircon grains in the Cranberry are rounded, therefore considered to be **detrital** (detritus means sediment) in origin, presenting an argument in favor of the Cranberry as a metamorphosed sedimentary rock.

A general consensus holds that the third oldest rock, the Beech Granite Gneiss, is a slightly metamorphosed pluton. The Beech appears to have a cross-cutting relationship with the Cranberry Gneiss and the Cloudland Gneiss—thus

Figure 8.3. Beech Granite Gneiss. The white blotches are actually phenocrysts of pink feldspar. The late Swiss alpine structural geologist Rudolf Trumpy, along with an unidentified student, are examining a cross-cutting dike of Bakersville gabbro. The dark-colored layer above them is another dike.

making it younger than both. The radiometric age of the Beech according to Su and others (1994) is 745 Ma. The composition of the Beech is typical granite—mainly quartz and feldspar with the pink orthoclase feldspar locally occurring as large crystals (**phenocrysts**) to form a porphyritic texture (see figure 8.3).

The youngest Neoproterozoic rock is the Bakersville Gabbro, or perhaps more properly, metagabbro, since it has been metamorphosed several times along with all the other Proterozoic rocks. The Bakersville is rather easily recognized as dark gray to greenish black dikes and sills that intrude the Cloudland, Cranberry, and Beech (see figure 8.4). Mineralogically it is composed mainly of ferromagnesian minerals (pyroxene), plagioclase feldspar, and magnetite. The white plagioclase often forms as phenocrysts in a porphyritic texture (see figure 8.5). The age of the Bakersville is 734 Ma (Goldberg and others 1986). It is possible that the intrusion of the Bakersville reflects early stages of rifting as the supercontinent of Rodinia began to break up.

The Grenville orogeny bonded many ancient pieces of the Earth's early crust to form the supercontinent Rodinia, including Laurentia (proto–North America)

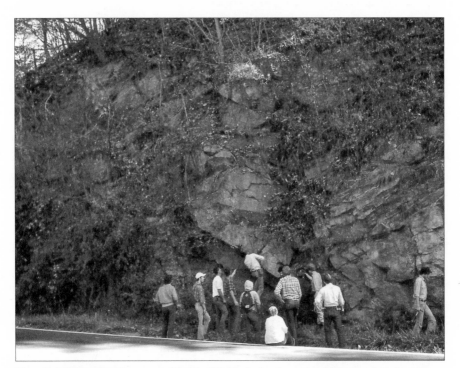

Figure 8.4. Bakersville dike cross-cutting the foliation in Cranberry Granite Gneiss. A student is pointing to the contact.

Figure 8.5. Porphyritic Bakersville Gabbro. The porphyritic texture indicates that the pluton cooled at different rates. The lighter-colored crystals are phenocrysts of plagio-clase feldspar that formed during a slow cooling rate.

and those that were later embraced by the future continents Gondwana and Laur-asia (continents associated with a later supercontinent—Pangea). The Neoprotero-zoic rocks of northeast Tennessee, including Cloudland, Cranberry, Beech, and Bakersville, were parts of a mountainous belt called the Grenville Mountains that evolved during various stages of the orogeny. Roots of these ancient mountains can be found throughout the Appalachian region from Newfoundland, Canada, to the Marathon region of Texas, where they once existed as a spectacular mountain chain along the margins of Rodinia. Subsequently, over millions of years, these early and middle Proterozoic mountains were leveled, and the eroded sediment was ultimately deposited to compile the sedimentary record of later Proterozoic and early Paleozoic time. Clasts of Cranberry/Beech can be actually identified in some conglomeratic rocks as young as the Ordovician (see figure 10.3). These are all the oldest rocks in Tennessee—Proterozoic, i.e., the deepest time recorded in the Tennessee's geologic history.

9

Sauk Sequence

> Nothing lasts long under the same form. I have seen what
> once was solid earth changed into sea, and lands created
> out of what was once ocean. Seashells lie far away from
> ocean's waves, and ancient anchors have been found on
> mountain tops.
>
> —Ovid, *Metamorphoses,* book 15

The deposition of the Sauk Sequence sedimentary rocks lasted for close to 300 million years from the Neoproterozoic Era up to the end of the Early Ordovician Epoch. The sedimentary rocks we see preserved today record the breakup of the supercontinent Rodinia and the subsequent formation of a new ocean basin called the Iapetus Ocean (in Greek mythology, Iapetus is the father of Atlantis). Erosion of the older Grenville Mountains on Rodinia had begun even as they were being uplifted, and it continued even as the eastern margin of Rodinia began to pull apart, about 750 million years ago. The rifting resulted in a series of deep rift basins along the trend of what is the present-day Appalachian Mountains. It also ultimately led to the separation of the proto–North American continent from Rodinia. As Iapetus was forming, the proto–North American plate is interpreted to have been located near the equator in a tropical to subtropical environment. However, at this same time, other parts of the world, some of which are presently tropical, were experiencing major glaciation. About 600 million years ago, as the Iapetus Ocean continued to spread, glaciers located on continents located on other plates began melting, resulting in a significant rise in sea level (see figure 7.4). A vast, shallow ocean began to spread from Iapetus across proto–North America, eroding and inundating all of what is presently Tennessee. With only a few minor breaks, sediment accumulated in this shallow sea from Neoproterozoic time through early Ordovician time. All the sedimentary rocks formed from these sediments make up the Sauk Sequence. Although rocks of the Sauk Sequence extend across Tennessee, major exposures at the land surface are restricted to the Unaka Mountains (Blue Ridge province) and the Valley and Ridge province of East Tennessee. Rocks of the Sauk Sequence lie deeply buried beneath younger rocks in Middle and West Tennessee.

Neoproterozoic rifting was not restricted to the eastern margin of Rodinia. Geophysicists have discovered the existence of a midcontinent rift system that includes the New Madrid rift complex that extends through the western part of Tennessee, where it is referred to as the Reelfoot rift. However, since proto–North America is still joined together in this region, at least for the time being, the midcontinent system can be considered a "failed" rift. Although Sauk Sequence rocks are not exposed in West Tennessee, they are deeply buried beneath younger rocks. Nevertheless, though hidden underneath Paleozoic, Mesozoic, and Cenozoic sediment, the New Madrid rift basin probably has influenced the location of the Mississippi River valley. The Mesozoic and Cenozoic sediments, the Zuni and Tejas sequences, in this rift zone, and, of course, its seismic activity (the New Madrid earthquakes—1811–1812) will be discussed later. Rocks of the Zuni and Tejas sequences are superimposed on the initial rift basin in the Mississippi Embayment and now make up the surficial deposits of West Tennessee.

The Sauk Sequence is time-transgressing (see figs. 7.2 and 7.4), thus in Tennessee the oldest rocks of the sequence are found in East Tennessee, where the initial sedimentary deposits related to the breakup of the supercontinent Rodinia occur. Extension of the crust during early stages of the pulling apart of Rodinia created numerous isolated rift basins that became recipients of sediment eroded from the nearby Grenville Mountains. Some of the faulting associated

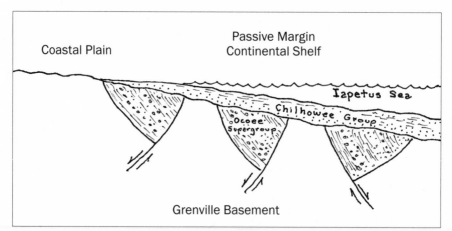

Figure 9.1. Cross-section sketch showing rift basins formed during the breakup of Rodinia and the coastal sediments later deposited on the resulting passive margin. The rift basins were filled with rocks of the Ocoee Supergroup and their equivalents in the Mount Rogers region and were later blanketed by coastal sediments on a passive continental margin as a sea began transgressing onto proto–North America.

with the rifting created conditions in the crust that provided avenues for magma to extend to the surface to form volcanoes or lava flows that became interlayered with the sediments being deposited at that time. One such basin with a volcanic history was the Mount Rogers Basin, located in northeastern Tennessee and adjacent areas of Virginia and North Carolina. In rift basins where faults did not penetrate deeply enough to cause magma to rise and form plutons, only the sediment eroded from adjacent highlands accumulated; for example, rocks exposed in Great Smoky Mountains National Park are formed solely from sediment deposited in the Ocoee Basin (see figure 9.1).

By around 550 Ma, as proto–North America drifted away from the other continental plates, the continental margin became passive, and the isolated rift basins were gradually blanketed by sediment deposited in a shallow seaway (the Appalachian Basin). The environment on this passive margin (tectonically quiescent) fluctuated between terrestrial and marine, mainly river and beach deposits (see figure 3.5). The Cambrian (earliest Paleozoic) Chilhowee Group that overlies rocks of both the Ocoee Supergroup in the Ocoee Basin and the Konnarock Formation in the Mount Rogers Basin represents the early stage of deposition on the passive margin (see figure 9.1). The source of the oldest Sauk (Proterozoic) sediment deposited in the initial rift basins were the rocks of the ancient Grenville Mountains; whereas the Paleozoic rocks, Cambrian through early Ordovician and deposited on the passive margin, consist of sediment largely eroded from the Precambrian rock interior of proto–North America—the **craton** (a relatively tectonically stable part of a continent) to the west.

The early Cambrian to late early Ordovician part of the Sauk Sequence is represented by sedimentary facies of a shallow transgressing sea that spread across almost all of proto–North America, including what is today recognized as Tennessee. The Cambrian rock units formed during this transgressing sequence include the Chilhowee Group, Shady Dolomite, the Rome Formation, the Conasauga Group, and the lowermost formation of the Knox Group—the Copper Ridge Dolomite/Conococheague Limestone—all of which are time-transgressing (progressively younger toward the west). The upper formations of the Knox Group are early Ordovician in age, and, at least in part, are likely associated with the regressive phase of the sea at the close of the Sauk Sequence.

Mount Rogers Basin

The best-known exposures of rocks formed in the Mount Rogers basin are in southwest Virginia near Whitetop Mountain and Mount Rogers near where Virginia is contiguous with Tennessee and North Carolina. The most detailed work done there is by Rankin (1993) and for this reason, even though correlative rocks occur in the very northeastern most corner of Johnson County, Tennessee, rock descriptions are mainly from rocks in Virginia. Good exposures of the rocks of the Mount Rogers Basin can be observed along the Virginia Creeper Trail, a "rails to trails" trail in Mount Rogers Recreational Area of Washington and Jefferson National Forest (http://vacreepertrail.com).

Rankin's (1993) interpretation of the Mount Rogers rocks divides the record into two formations. The older rock unit, the Mount Rogers Formation, is interpreted as mostly volcanic in origin, while the overlying Konnarock Formation is interpreted as being deposited in a glacial environment. The Mount Rogers Formation rests nonconformably (see figure 3.14C) upon the eroded surface of Grenville rocks, and rocks of the Chilhowee Group unconformably overlie the Konnarock Formation.

Mount Rogers Formation

According to Rankin, 50 to 60 percent of the approximately 6000 feet of the Mount Rogers Formation is an extrusive igneous rock called **rhyolite** (the aphanitic equivalent of granite (see figure 3.2) with the remainder consisting of in-

Figure 9.2. Diamictite (tillite) in the Konnarock Formation exposed along US Highway 58 east of Damascus in the George Washington and Jefferson National Forests, Virginia. The large clasts are rocks eroded from the Grenville Mountains.

terbedded clastic sedimentary rocks—conglomerate and sandstone. Most of the rhyolite is concentrated in three thick masses that Rankin interprets as volcanic centers. He estimated that between 200 and 400 square miles of rhyolite in the form of **lapilli** (walnut-sized pieces of solidified lava), pumice, **welded tuff** (volcanic ash bonded by heat), and lava flows were erupted from these centers. Based on these variations in texture and structure Rankin was prompted to subdivide the formation into nine members, including Welded Tuff Member, Lava Member, Lapilli Tuff Member, etc.

Konnarock Formation

The Konnarock Formation is the only currently known Neoproterozoic rock in Tennessee with an origin directly related to the cryosphere. According to Rankin, the Konnarock is on the order of several thousand feet in thickness and composed of maroon diamictite, rhythmite, and pink arkose. **Diamictite** is a poorly to non-sorted conglomerate with a wide range of clasts ranging from granule to boulder in size and supported by an argillaceous (clay) matrix; **rhythmites** are repetitions of graded sets or couplets of siltstone or fine-grained sandstone at the base and mudstone at the top; and **arkose** is sandstone composed mostly of feldspar grains.

The massively bedded diamictite of the Konnarock is interpreted as a glacial **till**—unstratified sediment deposited by glacial ice that has been lithified to form the rock **tillite** (Rankin 1993). The matrix-supported conglomerate (diamictite) contains clasts that range in size from clay to boulders more than three feet in diameter (see figure 9.2). Based on their composition, nearly 80 percent of the clasts were eroded from igneous or metamorphic rocks of the Grenville basement, with the balance made up of clasts derived from the Mount Rogers Formation.

One interpretation of the tillite is that it was formed by alpine (mountain) glaciers on the volcanic mountains of the volcanic centers of Mount Rogers Formation; albeit other evidence of glaciation, such as striations or gouging, on bedrock beneath the tillite caused by abrasion by the ice is apparently lacking.

The thin couplets of sediment of rhythmites are interpreted to reflect periodic changes in deposition over time (see figure 9.3). A **varve** is a rhythmite interpreted to be an annual lake (**lacustrine**) deposit seasonally controlled in a glacially related climate. During the summer, winds agitate lake water, holding fine mud in suspension and allowing coarse silt and sand to be deposited. Then, during winter, when the lake freezes, the fine suspended mud is allowed to settle, and no coarse sediment is carried to the lake. The Konnarock rhythmites are interpreted as varves. Some very interesting features associated with the Konnarock rhythmites are **dropstones** (clasts transported on ice rafts or small icebergs that fall into the lake sediment when the ice melts). Figure 9.4 shows a rhythmite with a dropstone. Most of the dropstones are very much like the clasts in the Mount Rogers tillites—granitic, and up to a few feet across.

Ocoee Basin

The rock record of the Ocoee Basin lies within eastern Tennessee, northern Georgia, and western North Carolina, an area that also embraces the Great Smoky Mountains National Park, straddling the line between eastern Tennessee and

Figure 9.3. Rhythmites or varves in the Konnarock Formation exposed along the Virginia Creeper Trail in the George Washington and Jefferson National Forests, Virginia.

Figure 9.4. A dropstone in a rhythmite exposed along the Virginia Creeper Trail, Washington and Jefferson National Forest, Virginia.

Figure 9.5. Bedding in the metasedimentary Thunderhead Sandstone along Hazel Creek, Great Smoky Mountains National Park, North Carolina. Note the round and elliptical dark areas scattered through the layers. These are nodules of calcareous material formed during deposition.

western North Carolina. Because of the complex structural relationships, i.e., folding and faulting, as well as dense vegetative cover and deep soils, there is no one location where a complete sequence of the nearly ten-mile thickness of rock formed in the Ocoee basin can be observed. However, the most recent interpretations by geologists have divided the thick mass into three groups. These three groups, from oldest to youngest, are Snowbird Group, Great Smoky Group, and Walden Creek Group. They constitute the Ocoee Supergroup. Although all of these rocks have been metamorphosed (low-grade), their sedimentary characteristics, such as bedding, have been preserved so well that they are commonly referred to as metasedimentary rocks (see figure 9.5). The makeup of the rocks reflects changes in environments caused by changing rates of subsidence resulting from continued rifting (plate divergence) of the basin (see figure 9.1). The fine-grained texture, thin-bedding, and other sedimentary structures of the rocks suggest times when the subsidence rate of the basin was slow and the depth of water relatively shallow. Graded bedding and scour structures that are formed by sediment abrading the surfaces of previously deposited sediment suggest that at other times subsidence was rapid and deposition was into a relatively deep basin. Beach or river environments have been suggested where the original sediment (premetamorphism) was clay, silt, or sand; turbidity currents interpreted where the sediment was coarse, graded, and large grains matrix supported; and anoxic (no oxygen) considered where the sediment was fine-grained, black, and containing pyrite or other iron-disulfide minerals. Figure 9.6 shows graded bedding in the Thunderhead Sandstone of the Great Smoky Group. Again, because the rocks have been intensely deformed, the terrain rugged, the vegetation dense, and the bedrock covered with deep soil or landslide debris, the interpretations of these rocks will continue long into the future.

Ocoee Supergroup

The Ocoee Supergroup nonconformably overlies Grenville basement rocks and is disconformably overlain by early Cambrian rocks. Rocks in all three groups contain little or no fossil record, and nowhere are all three (Snowbird, Great Smoky, Walden Creek) found in a contiguous sequence. The lack of a fossil record, undoubtedly related to the paleoenvironments in which these rocks formed, confounds the establishment of stratigraphic relationships. However, Snowbird is considered the oldest in that it can be observed to nonconformably rest upon

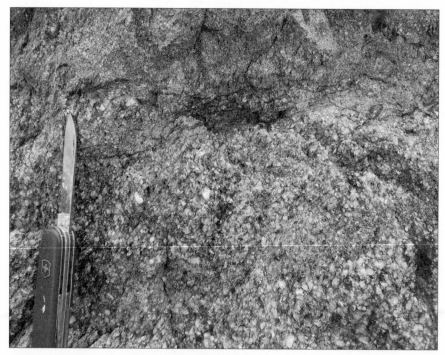

Figure 9.6. Graded bedding in the Thunderhead Sandstone.

crystalline Grenville basement rocks; and the Walden Creek Group is considered the youngest because it is observed to underlie the Cambrian Chilhowee Group. Unraveling the relationships among and within each of the three groups is confounded by the presence of major thrust faults that transect the Blue Ridge. One such dilemma is presented by the structural relationships of the Greenbriar thrust fault, perhaps the oldest fault in the Great Smoky Mountains. Present geologic maps of the Great Smoky Mountains National Park (King 1964 and King and others 1968) show the younger Great Smoky Group as thrust faulted over the older Snowbird Group—not what is expected in a typical thrust fault. Rast and Kohles (1986) proposed an attractive model to explain this relationship. Instead of a single Ocoee basin, there were two **grabens** (rift basins) separated by a **horst** (an uplifted block). One graben, which they call the Gatlinburg Basin, formed first and includes the sediment of the older Snowbird Group derived from the Grenville basement rocks. A horst, perhaps related to intrusion of granitic plutons, resulted in the formation of a second graben to the east called the Clingman Basin, which received mostly sediment, now forming the Great Smoky Group. It is possible that the Great Smoky Group is younger than the Snowbird or they could actually be at least in part the same age. Regardless, if the Greenbriar fault originated in the eastern Clingman Basin, the younger or same-age rocks of the Great Smoky Group could form the hanging wall and the Snowbird Group form the footwall of the Greenbriar fault.

Of course, the reconstruction of these basins and the horst is based solely on the stratigraphy of the units and their relationship to the Greenbriar fault—there are no vestiges of the basins or the horst—both presumably being part of the footwall of the Greenbriar fault. Strata of the Walden Creek Group represent the ultimate filling of the Gatlinburg basin. All the rocks of the Ocoee Supergroup contain evidence of metamorphic zones from chlorite on the west through garnet on the east (see chapter 3), yet they retain sedimentary characteristics; therefore, they are often referred to as **metasedimentary**.

Snowbird Group

Five formations are included in the Snowbird Group: the Wading Branch Formation, the Long Arm Quartzite, the Metcalf Phyllite, the Roaring Fork Sandstone, and the Pigeon Siltstone. The Wading Branch, estimated to be 1500 feet thick, is mostly sandy slate to pebbly feldspathic sandstone that has been interpreted to represent alluvial (stream) deposits formed in the early stage of rift basin formation (Gatlinburg Basin?). As the basin deepened, the Long Arm, about 5000 feet thick and consisting of mostly cross-bedded sediment rich in feldspar (arkose), the Roaring Fork, up to 7000 feet thick and mostly siltstone and feldspar-rich sandstone, and the Pigeon, mostly siltstone and up to 10,000 feet thick, were deposited on floodplains of streams that formed a deltaic system. The Metcalf, now a phyllite up to 5000 feet in thickness, consists of fine-grained clay and silt that was carried offshore to the mid-portions of the basin.

Great Smoky Group

Although other stratigraphic names have been applied to the rocks of the Great Smoky Group in other parts of the Blue Ridge province in Great Smoky Mountains National Park region, the three formations normally recognized are the Elkmont Sandstone, the Thunderhead Sandstone, and the Anakeesta Formation. This group is dominated by massive layers of coarse graywacke (metagraywacke) and arkose and is estimated to range in thickness between 14,000 and 25,000 feet. All three units probably intertongue.

The Elkmont Sandstone is coarse- to fine-grained gray arkose, graywacke, and fine conglomerate. In general, it is much finer in texture than the overlying Thunderhead. The Elkmont is estimated to be between 1000 and 8000 feet in thickness. Conformably overlying the Elkmont, and possibly grading into it or intertonguing with it, is the Thunderhead Sandstone. The Thunderhead consists of coarse, gray feldspathic sandstone, graywacke, and conglomerate that occur in massive beds typically forming topographic ledges, cliffs, and waterfalls. Graded-bedding and clasts of partially lithified rocks, typical of turbidite deposits, are common in the Thunderhead (see figure 9.6). The turbidity currents resulted from erosion of the uplifted horsts of Grenville rocks during the initial breakup of Rodinia. The presence of blue quartz is very characteristic of the Thunderhead and suggests that the source for the sediments was a crystalline rock such as granite. With a thickness between 5500 and 6500 feet, the Thunderhead forms the core of the Great Smoky Mountains.

Overlying the Thunderhead and capping the crests of many of the ridges and mountains in the Great Smokies, such as Mount LeConte, Clingmans Dome, and the Chimney Tops, is the Anakeesta Formation—a dark to bluish-gray to black slate or phyllite with interbeds of dark gray, fine-grained sandstone (see figure 9.7). Because of the black color and the presence of the mineral pyrite, the Anakeesta is believed to have been deposited in deep, oxygen-deprived, **euxinic** (poor circulation) pockets of the basins. When weathered, the pyrite oxidizes and forms various sulfo-salt minerals and sulfuric acid. The Alum Cave Bluff in Great Smoky Mountains National Park, once mined for Epsom salts (various sulfo-salt minerals), displays many of the products of chemical weathering of the Anakeesta (see figure 9.8). When the Anakeesta is excavated by human activities or broken up in natural landslides, steps have to be taken to protect aquatic environments from acidic drainage. The Anakeesta is estimated to be 3000 to 4500 feet thick.

Figure 9.7. The Chimney Tops, Great Smoky Mountains National Park. Typical topography formed by weathering and erosion of the Anakeesta Formation. Landslides are common on ridges underlain by the Anakeesta.

Walden Creek Group

The Walden Creek Group underlies the foothills of the Blue Ridge or Smoky Mountains. Lithologically the group consists of a heterogeneous succession of shale and siltstone with sandstone and conglomerate in discontinuous masses. Carbonate rocks such as limestone and dolomite are also present in minor amounts—often as large masses of **breccia** (a conglomerate composed of angular-shaped clasts). Various geologists have subdivided the group into formations; however, for all intents and purposes, it can be recognized as a single unit of sandstones, shales, and conglomerates about 8000 feet thick (deep) that culminated the filling of the Ocoee Basin.

The Passive Margin and the Epicontinental Sea

As the subsidence of the Ocoee Basin, Mount Rogers Basin, and other rift basins along the margin of the proto–North American continent began to diminish, the major rift to the east continued, causing the eastern edge of the proto–North American continent to become a passive margin analogous to the present-day Atlantic coast. The continuous spreading of the rift, plus the melting of glaciers in other parts of the world, caused the Iapetus Ocean to evolve into a shallow sea that slowly transgressed across proto–North America—first blanketing the older rocks formed in the Ocoee and Mount Rogers rift basins with beach sand (see figs. 3.5 and 9.1), then spreading across proto–North America to form a vast epicontinental sea. These early sediments formed the Chilhowee Group and mark the beginning of the Paleozoic Era, 542 Ma.

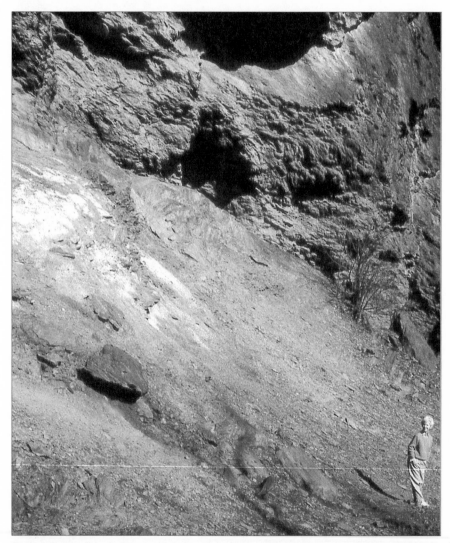

Figure 9.8. Alum Cave Bluff in the Anakeesta Formation, on the trail to Mount LeConte in Great Smoky Mountains National Park. The white area in the upper left corner is a patch of various efflorescent sulfo-salt minerals formed by the chemical weathering of the mineral pyrite.

Chilhowee Group

Sediment of the early Cambrian Chilhowee Group was deposited along a passive continental margin that formed the ancient coast of proto–North America. It has not been unequivocally proven, but the Chilhowee Group presumably rests unconformably upon the Neoproterozoic Walden Creek Group. Prominent ridges in East Tennessee such as Chilhowee Mountain, Starr Mountain, Bean Mountain, Holston Mountain, Iron Mountain, and Buffalo Mountain are formed by resistant quartzites of the Chilhowee Group and typically mark the boundary between the Blue Ridge and the Valley and Ridge provinces. At this boundary in Tennessee, the Chilhowee rocks of the Blue Ridge province occur at the surface where they have been thrust faulted onto younger Paleozoic rocks of the Valley and Ridge province.

Although no unconformities are apparent within the group, it is divisible into conformable formations—three units in northeastern Tennessee where they are well exposed in the Doe River gorge in Carter County. From oldest to youngest they are the Unicoi Formation (2000–5000 feet thick), the Hampton Formation (500–2000 feet thick), and the Erwin Formation (1000–1500 feet thick). Elsewhere in East Tennessee, five formations, from oldest to youngest, compose the Chilhowee Group: Cochran Conglomerate (~1200 feet thick), Nichols Shale (~700 feet thick), Nebo Sandstone (250 feet thick), Murray Shale (~500 feet thick), and Hesse Sandstone (~600 feet thick). The Hesse, Murray, and Nebo sequence of formations are found mainly south of the French Broad River and probably correlate with the Erwin in northeastern Tennessee; the Nichols, south of the French Broad, correlates with the Hampton to the north, and the Unicoi in the north correlates with the Cochran to the south.

As sediment gets cycled in the rock cycle again and again—that is, as it gets weathered, eroded, transported, and deposited over and over again throughout geologic time—the feldspars and ferromagnesian minerals become clay, and the quartz remains quartz but becomes rounded and sorted. The resistant mineral quartz dominates sediment of the Chilhowee Group, having been derived mainly from older Neoproterozoic Ocoee clastic sedimentary rocks that in turn had been derived from older Proterozoic Grenville crystalline rocks. The sandstone units are composed of very resistant vitreous quartzite, and the shales are typically greenish in color and micaceous.

Although the Unicoi Formation in northeastern Tennessee is a sequence of feldspathic sandstone, conglomerate, graywacke, siltstone, and shale, it differs from its correlative, the Cochran Conglomerate to the south, mainly by containing several layers of **amygdaloidal basalt** (named such because of the *amygdules,* almond-shaped gas holes in the basalt filled with secondary minerals such as fluorite) preserved near the middle and bottom of the formation. Because of the presence of amygdules, one explanation is that these are lava flows rather than plutons and are related to the volcanic history of the nearby Mount Rogers basin noted above.

Despite the fact that the Chilhowee sediments accumulated in a relatively high-energy beach environment of waves and surf, the rocks contain evidence of early life. Outcrops of the Nebo Sandstone near Walland, Tennessee, contain the trace fossils of *Scolithus linearis,* vertical worm borrows (see figure 9.9). Also, ostracodes (a primitive arthropod) have been described in the Murray Shale near Murray Gap on Chilhowee Mountain, and trace fossils have been reported in the Nichols Shale.

Shady Dolomite

Overlying the Chilhowee is the Shady Dolomite, which only occurs at the surface in eastern portions of the Valley and Ridge province; however, even there, because of weathering, outcrops are scarce. Typically residual yellowish-brown clays with "jasperoid" (a form of reddish chert) are virtually all that suggest the presence of a formation nearly 1000 feet thick.

Figure 9.9 Scolithus linearis (worm burrows) in the Nebo Sandstone on Chilhowee Mountain near Walland, Tennessee. Note the vertical tubes in the layer above the pocketknife. The round features on the surfaces of bedding planes are the tops of worm tubes.

One possible scenario is that the Shady was deposited as a carbonate bank or reef along the eastern hinge of a gradually subsiding Appalachian basin. Sediment-free water offshore from the beach, where sands of the Chilhowee were shifting with the tides, was a suitable environment for cyanobacteria and other organisms to accumulate in the form of an embankment, much like a reef. As the sea level continued to rise (or the land subsided), a shallow sea transgressed westward onto the **craton** (stable interior of the continent), and the carbonate bank migrated westward with the transgression in order to maintain the proper energy from sunlight for the organisms to survive.

Rome Formation

After a few tens of millions of years, the transgressing sea had overlapped the worn-down Grenville Mountains and rift basins (Ocoee, Mount Rogers, and others) and moved toward the interior of the continent (the craton). During this time the crystalline, granitoid bedrock of the craton was being weathered and eroded and, given the proximity to the equator at that time, was possibly being blanketed with a thick layer of reddish residuum, perhaps a **laterite** soil (a present-day soil type associated with tropical environments). The reddish mud was washed from the craton into the transgressing sea and soon mixed with carbonates of the Shady to form a gradational contact between the Shady and the red-colored Rome Formation. The Rome Formation is actually the western facies of the Shady Dolomite.

The Rome was deposited in a shallow tidal-flat environment between the craton and the carbonate bank that was sometimes inundated by shallow water (subtidal) and at times subaerially exposed (supratidal). The occurrence of many ripple marks is evidence of subtidal conditions and the presence of salt (halite) hoppers, desiccation or mud cracks, raindrop impressions, and small crystalline cavities called vugs that may be pseudomorphs of other evaporite minerals, such as gypsum, suggest periods when the tidal flat was high and dry.

The total thickness of the Rome may reach 2000 feet. However, as will be discussed later, the Rome plays an important role in the structure of the Valley and Ridge province, and the lower portion of the Rome is often occluded by later thrust faulting. The great thickness of the Rome and the evidence of a shallow-water depositional environment seem contradictory, but it must be remembered that deposition was taking place over a very long period of time, and, at the same time, the Appalachian basin was very slowly subsiding. Variegated (red, green, yellow) shales and siltstones are characteristic of the formation, but gray, fine-grained sandstone and limestone and dolomite also are present, though in minor amounts. Limestone and dolomite occur mostly in the lower Rome,

Figure 9.10. The trilobite *Olenellus* is a body fossil (actual remains) that is recognized as a guide or index fossil for the Cambrian Period. Trilobites were scavengers that primarily inhabited the seafloor. This trilobite was collected near Sneedville, Tennessee.

especially in the east (near the carbonate bank), and sandstone is thicker and more prevalent to the west near the sediment source and the shore of the advancing sea. Although the sediment deposited along the shore of the advancing sea are dominantly siliceous, carbonate cement is ubiquitous in most of the rocks as a result of the composition of the seawater circulating in the sediment. Trace fossils including various trails and tracks are abundant in the Rome, suggesting significant benthic (bottom-dwelling) life. The trilobite *Olenellus* (see figure 9.10), though rare, is the typical body fossil in the Rome and was probably the creature making many of the trails and tracks (see figure 9.11).

The Rome Formation is the oldest one exposed in the Valley and Ridge province of East Tennessee. As will be discussed in a later chapter, the Rome is the **décollement** (glide zone) fault along which the pile of sedimentary rocks deposited in the Appalachian basin were stripped from basement rocks and pushed westward during the Alleghanian orogeny. During this faulting, numerous slices of strata bounded by the Rome on the bottom were thrust to the surface in imbricate fashion like shingles on a roof, repeating the stratigraphic section over and

Figure 9.11. Trace fossils (trails and tracks formed by organisms) in the Rome Formation suggesting the presence of critters crawling around on the bottom of the shallow Rome sea.

over again (see figs. 1.7 and 13.2). Other than in the Valley and Ridge province in the east, the Rome occurs in the subsurface across the remainder of the state.

Conasauga Group

This group of formations provides a classic example of stratigraphic facies (see chapter 7). Throughout the Cambrian Period, the shallow sea continued to transgress across Tennessee as the Appalachian Basin continued to subside. The early Cambrian Rome formation is a time-transgressing unit that, like all formations deposited in an advancing sea, grades upward into the younger overlying Cambrian Conasauga Group that on the basis of fossil evidence is classified as middle-late Cambrian in age. As manifest by the formations of the Conasauga Group, the depositional environment remains about the same as conditions during Shady and Rome times—a carbonate bank to the east, terrigenous clastics being washed in from the craton lying to the west, and an intertonguing of carbonate sediments and terrigenous clastics in the middle. Figure 7.3 shows the facies of the nearly 2000-foot-thick Conasauga Group deposited in the continually subsiding Appalachian Basin.

The Honaker Dolomite is a good example of carbonate rock dominance in the eastern phase of the Conasauga. It is especially prevalent in northeastern Tennessee, where it is a dark gray medium-bedded dolomite with dark limestone beds. Its organic content causes the Honaker to give off a characteristic fetid (rotten egg) odor when struck with a hammer. Stromatolite-forming cyanobacteria were major contributors to the accumulation of the Honaker sediment and are probably responsible for the organic content (see figure 9.12). The total thickness of the Honaker is estimated to be about 1500 feet. Also present in the eastern phase is a thin shale, on the average about 100 feet thick; this is the Nolichucky Shale lying above the Honaker and separating it from the Maynardville Limestone, which may be 150 to 400 feet thick.

The middle phase of the Conasauga, where the facies intertongue, consists of six formations of alternating limestone and shale (see figure 7.3). The lowermost unit is the Pumpkin Valley Shale (100 to 600 feet thick), which resembles, for the most part, the variegated beds of the underlying Rome Formation. Overlying the Pumpkin Valley is the Rutledge Limestone (100 to 500 feet thick). Above it is the thin Rogersville Shale (typically 25 to 80 feet thick), followed by the Maryville Limestone. The dark gray to black Maryville (300 to 800 feet thick) has periodically been quarried as dimension stone near the community of Thorn Hill, Tennessee, and marketed as Imperial Black Marble. Above the Maryville is the Nolichucky Shale (up to 900 feet thick in middle phase). The trilobite fauna of the Nolichucky, the one unit of the Conasauga that appears in all three phases, is regarded as Late Cambrian in age. And, lastly, the uppermost unit of the Conasauga is the Maynardville Limestone (150 to 400 feet thick). The boundary between the Maynardville and the overlying Copper Ridge Dolomite of the Knox Group is hard to recognize; however, in most cases the Maynardville is distinguishable from the overlying Knox by its paucity of chert when compared to Knox units.

The western exposures of the Conasauga in the Valley and Ridge province are dominated by shale, and the group is difficult to subdivide into formations. Although the prevalence of shale in this part of the formation is a result of proximity to the western cratonic sediment source, the fact that the craton was being depleted of sediment is manifest in the greater abundance of carbonate rocks in the Conasauga when compared to the underlying (older) Rome Formation. This increase in carbonate rocks is also a reflection on the continually transgressing sea across Tennessee and the rest of the proto–North American continent. Like the underlying Rome Formation, Conasauga rocks occur at the surface only in East Tennessee.

Knox Group

The shallow transgressing sea eventually succeeded in inundating all of Tennessee and much of what is now North America. A broad carbonate platform nearly 3000 feet thick was built primarily by cyanobacteria and other organisms. A modern analog of this environment is that of the Bahamas, where carbonate sediment is presently being biochemically deposited. This environment persisted during late Cambrian and early Ordovician and marks the end of the Sauk Sequence.

The Knox Group has two distinct facies—an eastern limestone-dominated facies and a northwestern dolomite dominated facies. The true paleogeographical relationship between the two facies cannot be discerned because the Pulaski thrust fault, a major fault in the Valley and Ridge province, has shortened the spatial relationship between the two facies. The northwestern phase may be interpreted as a product of **dolomitization** of limestone (changing limestone into dolomite by the addition of magnesium). The sediment in the west accumulated in a tidal flat environment with abundant stromatolitic buildups of cyanobacteria (reefs). Evaporation in the tidal flat environment resulted in magnesium replacing calcium to form dolomite. Nearer to the open sea to the east, seawater circulated more freely, eliminating the process of evaporation.

In the northwestern parts of East Tennessee, the Knox has been divided into five dominantly cherty, dolomitic formations: the Copper Ridge (about 1000 feet thick), the Chepultepec Dolomite (about 800 feet thick), the Longview Dolomite (about 300 feet thick), the Kingsport Formation (about 250 feet thick), and the Mascot Dolomite (about 800 feet thick). The bottom of a dolomitic sandstone at the base of the Chepultepec that is interpreted by some geologists to be an **eolian** (wind-blown) deposit marks the time boundary between the Cambrian and Ordovician Periods. All these formations are characterized by an abundance of buff-colored chert in a reddish-orange-yellow residuum that blankets the bedrock. The chert comes in many sizes and shapes and often contains fossils. As the clay-rich residuum is gradually eroded away, the chert remains to form an armor that ultimately retards erosion to the extent that these formations typically underlie broad ridges with the very cherty Copper Ridge and Longview formations that produce prodigious amounts of chert, forming the two highest portions of the ridge. Much of the chert consists of silicified stromatolitic-forming cyanobacteria that were instrumental in the accumulation of the carbonate sediment (see figure 9.12).

The eastern, less dolomitic phase of the Knox, consisting of relatively chert-free limestone, is dominant in the northeastern part of Tennessee. The Knox is comprised of two formations in this phase: the Conococheague Limestone (about

Figure 9.12. Stromatolites in the Knox Group formed during deposition by cyannobacteria. (Photograph courtesy of Harry Moore.)

1000 feet thick) and the Jonesboro Limestone (about 2000 feet thick). Except for sandstone in the basal portion of the Jonesboro, the formations are rather difficult to distinguish.

Exposures of the Knox are widespread in East Tennessee, but, for the most part, the unit occurs only in the subsurface across the remainder of the state. Exceptions—where the Knox occurs as outcrops west of the Valley and Ridge province of East Tennessee—include the core of the faulted anticline in Sequatchie Valley, in Bledsoe, Sequatchie, and Marion Counties, and in two meteorite-impact structures, Flynn Creek in Jackson County and Wells Creek in Stewart County.

About 475 Ma, the shallow sea once covering the carbonate platform began to diminish as a result of uplifting to the east, probably initiated by the collision of small terranes with proto–North America as the Iapetus Ocean began closing. As sea level dropped, the carbonate platform was gradually exposed. The carbonate rocks of the Knox Group forming the surface of the newly exposed land were

Figure 9.13. Collapse breccia at the unconformity at the top of the Knox Group dividing the Sauk Sequence from the Tippecanoe Sequence. The breccia resulted from the collapse of bedrock and chert fragments into a sinkhole. This is an example of paleokarst (ancient karst). The dark angular fragments are chert.

weathered and eroded to form a karst replete with sinkholes, subsurface solution channels, and caves. Because the eroded karst surface contained collapsed sinkholes and a regolith of chert fragments, the paleokarst (ancient karst) surface is recognized today by breccia formed by the collapse of the sinkholes formed during early Ordovician time and layers of cherty conglomerate in the overlying layers of the Tippecanoe Sequence (see figure 9.13). The erosional event marks the close of the Sauk Sequence.

10

Tippecanoe Sequence

> Since things are far more ancient than letters, it is not to
> be wondered at if in our days there exists no record of how
> the aforesaid seas extended over so many countries.... But
> sufficient for us is the testimony of things produced in the
> salt waters and now found again in the high mountains,
> sometimes at a distance from the seas.
>
> —Leonardo da Vinci, *Notebooks* (1508–1518)

Middle Ordovician

About 475 million years ago, or so, convection currents within the Earth's mantle changed, causing the Iapetus Ocean to begin closing, with the plate supporting Gondwana moving toward the southeastern border of the proto–North American plate. In the process, several relatively small terranes collided with the eastern edge of proto–North America, where they were subducted. These collisions created profound changes in the landscape of proto–North America and are well documented in the rock record of Tennessee—namely, the environments of sites being eroded and those where sediment was deposited. The rock record reveals three major changes in the events of Tennessee's geologic past: first, as the Iapetus Ocean began to close during plate convergence, crust was progressively down warped, forming a foredeep basin with a westerly hinge line in what is now East Tennessee; second, the main provenance (source) for sediment forming its bedrock shifted from the craton in the west to a mountainous area forming in the east by terranes colliding and subducting beneath proto–North America; and, third, there is evidence for volcanism found in the sedimentary record (see figure 10.1).

When tracing the history of geologic mapping in Tennessee, it can be noted that rocks of the middle Ordovician section have received a plethora of formational names. At present, these different formations, many of which are correlative, are recognized as the Chickamauga Group. The application of numerous

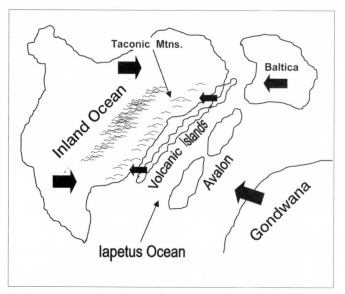

Figure 10.1. The closing of the Iapetus Ocean and the development of a foredeep basin in the Appalachian region. Note the presence of volcanic islands off the east coast of proto–North America—probable source for the middle Ordovician bentonite (altered volcanic ash) deposits (see fig. 10.2).

names to this stratigraphic interval is a reflection of the array of depositional environments (facies) that include the dominant clastic sediments of the actively subsiding foredeep basin in the east and the carbonate muds and sands of tidal flats that extended westward to what is now the Central Basin (Nashville) of Tennessee. The thicknesses for all of the units vary depending on where in the depositional model the sediment accumulated. There are two main environments in the model—the foredeep basin containing a thick sequence of strata dominated by clastics, and a shallow **epeiric** (epicontinental) sea that flooded the interior of the North American continent beyond the western hinge of the foredeep to form a vast "carbonate bank." Correlation of rock units that make up the middle Ordovician record is enhanced by the presence of at least four or maybe five ancient volcanic ash deposits (now altered to bentonite clay). Each volcanic eruption from volcanic arcs associated with terranes colliding with proto–North America along the eastern margin of the continent spread ash over the various depositional environments that existed across Tennessee. Thus, each layer of altered volcanic ash is a time line regardless of whether it was deposited in the subsiding foredeep basin or on the carbonate bank.

Figure 10.2 is a model proposed by Walker and others (1980) showing the depositional environments in the eastern part of Tennessee during middle Or-

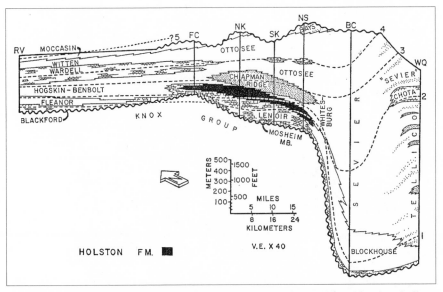

Figure 10.2. Cross-section from Raccoon Valley (line RV) on the west to Wildwood Quadrangle (line WQ, Tennessee) on the east showing facies of the Chickamauga Group in East Tennessee. The vertical lines RV, NK, SK, NS, BC, and WQ represent the locations of the measured sections of strata. The numbered dashed lines represent the layers of bentonite (altered volcanic ash) that enable the correlation of time across the area (Walker, Broadhead, Keller 1980).

dovician time. Based on its shape, which resembles a gun, it is affectionately referred to by Tennessee geologists as the "pistol diagram." The section line BC in figure 10.2 shows the great thickness of the sediment that accumulated during middle Ordovician time in a relatively rapidly subsiding foredeep basin—perhaps as much as 8000 feet. The rocks resulting from this basinal depositional environment are mostly dark gray shales and calcareous siltstones. Intercalated with the vast thickness of shales and siltstones are beds of sandstone and siltstone, mostly as turbidites, sloughed off the flanks of mountains rising to the east to form interbedded units along the eastern edge of the basin. The shales, though dark when fresh, turn a buff color when weathered, which is their most common appearance. Some conglomeratic turbidites, such as those exposed below South Holston Dam in Sullivan County, Tennessee, contain clasts of nearly every older rock previously deposited, including clasts of crystalline basement rocks (see figure 10.3). The dominant fauna of the foredeep facies consists of **graptolites** (see figure 10.4) that were able to survive in the deep water because they were floaters. Upon death, their remains could be preserved in the quiet, low oxygenated water on the basin bottom where scavenging predators were absent.

The western hinge (section lines NK and SK of figure 10.2) of the basin is carbonate-rich, as it is supported by a thriving reef complex composed of life

Figure 10.3. Turbidite conglomerate deposit composed of clasts eroded from Precambrian through Ordovician rocks on the eastern margin of the deep basin shown in fig. 10.2. Photograph taken of outcrop below TVA South Holston Dam.

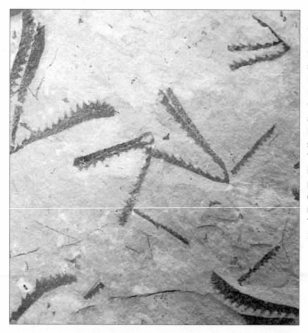

Figure 10.4. Graptolites are about the only fossils found in the middle Ordovician Sevier Shale. Graptolites are soft-bodied floating forms of life. When they die, their remains settle to the seafloor where they are preserved as films of carbon in the bottom sediments of the subsiding foredeep basin. Their remains likely stayed intact because water depth precluded the presence of scavenging critters who might have consumed them. The average length is about 1 inch.

nourished by sunlight afforded by relatively shallow water and nutrients washed in from the open water to the east. This formation is the Holston Formation (referred to most often as the Holston "marble"). The Holston is coarsely crystalline limestone ranging in thickness from about 200 to 600 feet. Depending on the amount of iron oxide present in the rock, the color of the Holston can be gray, light to dark pink, or "cedar" red. Because it polishes well and has low permeability, it has been widely quarried and used as dimension stone in various architectural renderings throughout the United States, including on many monuments and buildings in Washington, DC (see chapter 16).

The ancient environment west of the Holston reef and extending on toward what is now the Nashville Basin was a carbonate shelf beneath a shallow sea containing small patch reefs, in many respects similar to the Holston, and skeletal carbonate sand banks eroded from the reefs. The rocks of this carbonate bank have been mapped and described by many different geologists over many years, and, as a result, various names have been applied to the stratigraphic units composing this interval of geologic history.

Geologists in the Nashville region have subdivided the dominantly limestone formations into two groups: a lower Stones River Group consisting of the Pierce/Murfreesboro Limestone, the Ridley Limestone, the Lebanon Limestone, and the Carters Limestone; and an upper Nashville Group comprising the Hermitage Formation, the Bigby-Cannon Limestone, and the Catheys Formation.

In East Tennessee, complex facies relationships are more prevalent than in the Nashville area, and many names have been applied to the rock units that are now recognized as the Chickamauga Group—in many cases different names were used for the same or correlative unit (see figure 10.2). Furthermore, direct correlation of the middle Ordovician rocks in East Tennessee with those in the Nashville area is difficult because the rocks cannot be traced on the surface between the two regions. Between East Tennessee and Nashville lies the Cumberland Plateau, where the middle Ordovician rocks occur in the subsurface. However, one can get a glimpse of the middle Ordovician rocks at the surface, where erosion of the Sequatchie anticline has exposed them in Sequatchie Valley on the Cumberland Plateau. Milici (1970a, 1970b) used Nashville stratigraphic nomenclature in describing the Sequatchie Valley rocks. Despite the problem of trying to correlate the facies based on physical parameters, the time equivalency of the middle Ordovician rocks in East Tennessee with those in Middle Tennessee is facilitated by the presence of the five distinct layers of **bentonite** (altered volcanic ash) that were spread penecontemporaneously upon the sediments as they were accumulating across Tennessee on the vast carbonate bank.

The Ordovician strata of Tennessee are representative of the "Great Ordovician Biodiversity Event" (GOBE) that might be the most significant and sustained increase in marine biodiversity in Earth's history. This condition is likely a result of a rather extraordinary paleogeographic setting around the world. Vast shallow continental margins in warm climates were brought about by the great dispersal of continents and high sea levels that resulted as supercontinent Rodinia continued to break up. Other contributory factors for this biotic explosion may have included high nutrient input to the oceans from volcanic activity and/or possibly to some degree asteroid impacts (Servais and others 2009). This abundance and diversity of marine fossils can be observed in Tennessee rocks ranging from middle Ordovician through Devonian in age. An example of some of the more common organisms found as fossils in Ordovician rocks of Tennessee are shown in figure 10.5. Figure 10.6 is a sample of Ordovician limestone composed mainly of fossil fragments much like the coquina limestone found along the Florida coast.

The open sea, from time to time, was restricted in the region now occupied by the Nashville Basin (currently a topographic low area). The Nashville region was periodically domed upward along an axis that extended through central Kentucky and Cincinnati, Ohio. Although this uplift could possibly be related in some way to the Taconic orogeny (terrane collisions to the east), it is more likely

Figure 10.5. A typical middle Ordovician seafloor with coral, swimming cephalopods, bryozoa, sea lilies, and a scavenging trilobite roaming the seafloor. (Exhibit by Chase Studios, Inc. Courtesy of the McClung Museum, University of Tennessee, Knoxville.)

Figure 10.6. A typical slab of fossiliferous middle Ordovician limestone. Such fossiliferous beds resemble the modern coquina limestone of St. Augustine, Florida. The assortment of fossils were probably brought together by tidal action. Among the obvious fossils are brachiopods (ribbed sea shell–looking), bryozoa (similar to chicken bones), pelmatazoa (round columns or beadlike pieces), and coral.

related to faulting in the subsurface basement of Grenville age rocks. The total up-arched region is called the Cincinnati Arch with the Tennessee portion being called the Nashville Dome (now, of course, eroded to form the Nashville Basin). The exact influence that the Nashville Dome had on Middle Tennessee stratigraphy is equivocal. When the dome area was below sea level (either sea level rose or the land sank), sediment would accumulate across the dome area, but when lifted above sea level (either sea level dropped or the land rose), it served as a sediment source by causing erosion of sediment that had been previously deposited on its summit, giving the impression that some sediments were never deposited or that the dome might have existed as an island with sediment accumulating around its periphery. **Isopach maps** (showing the thickness of rock units) indicate thinning of units around the periphery of the dome area but, for all intents and purposes, also suggest that the dome was submerged during most of middle Ordovician time (Wilson 1949). It is not obvious whether the sea-level relationship with the dome resulted from the up or down movement of the dome (tectonic cause) or the rise and fall of sea level (eustatic cause). It was probably some of both. In general, Ordovician rocks become thinner and more carbonate in composition westward from eastern Tennessee toward central Tennessee.

Late Ordovician

The same tectonic forces that initiated the foredeep basin in East Tennessee escalated throughout middle Ordovician time until, by late Ordovician time, a significant mountain chain existed along the eastern margin of the Appalachian basin. The mountain-building event is known as the **Taconic orogeny**, and the mountains created are referred to as the **Taconic Mountains**. Clastic sediment shed from this mountainous terrain topped off the foredeep basin and began to muddy the waters across the carbonate platform to the west. In East Tennessee, the Moccasin Formation (mostly maroon calcareous shale 800 to 1000 feet thick) and the Bays Formation (mostly maroon siltstone and sandstone 1000 feet thick) form the facies nearest to the rising mountains (see figure 10.12). Bays Mountain, a prominent mountain in upper East Tennessee, is underlain by sandstone of the Bays Formation. Other late Ordovician facies in East Tennessee include the Martinsburg Formation (medium- to dark-gray shale and limestone about 1800 feet thick), the Juniata Formation (maroon siltstone and sandstone, about 360 feet thick), and the Sequatchie Formation (maroon and gray shaly limestone, mottled greenish, 200 feet thick). The Juniata and Sequatchie overlie the Martinsburg and are interpreted as lateral equivalents, the former representing a sandy and silty tidal flat environment to the north and east, and the later a more open marine environment, including more calcareous rocks extending to the south and west.

Late Ordovician rocks west of the Appalachian Basin are predominantly limestone, deposited in tidal flat environments much the same as the Sequatchie Formation and the underlying middle Ordovician rocks. The influence of the Nashville Dome on the depositional environment during this time continues to be equivocal; however, the fact that locally there is complete absence of formations such as the Inman Formation (limestone and shale, 0 to 50 feet thick), the Leipers Formation (a shaly limestone, 0 to 150 feet thick), and Sequatchie Formation (calcareous shale and shaly limestone, 0 to 165 feet thick) suggests the presence of an emergent dome in the present Nashville region that was either being eroded or just not receiving sediment.

Silurian

Silurian rocks in Tennessee differ vastly from east to west. The eastern portion of Tennessee was proximal to the influence of the Taconic Mountains formed during the Taconic orogeny; thus the foreland basin that formed on the west flank of the Taconic Mountains received a significant volume of clastic sediments. The Clinch Sandstone (about 700 feet thick) is a white, well-sorted sandstone interpreted to represent an ancient beach along this basin margin. The Clinch, a significant ridge/mountain maker in Tennessee (Clinch Mountain/House Mountain), and its equivalent, the ridge-forming Tuscarora Conglomerate to the north, form a near continuous ridge extending southward to about the latitude of Knoxville. This was a formidable barrier to westward migration of early pioneers. Sedimentary structures such as cross-bedding suggest a northeastern source for these mountain-forming sandstones and conglomerates. The lateral equivalent of the

Clinch Formation to the south, farther from the influence of the Taconic Mountains, is the Rockwood Formation (200 to 800 feet thick). The distinguishing lithology of the Rockwood is calcareous hematitic sandstone, the beds of which range in thickness from inches to several feet. The maroon hematitic beds are **oölitic** and fossiliferous. Rock with an oölitic texture contain small, spherical BB-sized grains usually formed by the precipitation of calcite, silica, or hematite in concentric bands around nuclei such as quartz grains. The oölitic texture and fossil content suggest that this part of the Silurian record was deposited in the tidal zone of the sea. The Rockwood was prospected for and locally mined as iron ore, mainly during the early twentieth century. The remainder of the Rockwood is gray siltstone and shale (see chapter 16). With the exception of a few localities such as Whiteoak Mountain in Hamilton and Bradley Counties, Tennessee, outcrops of the Rockwood, because of structural relationships, occur mainly along the eastern margin (escarpment) of the Cumberland Plateau.

The Rockwood grades westwardly into the Brassfield Formation, where in Sequatchie Valley it is mostly an olive gray calcareous shale, 60 to 130 feet thick. However, farther west, the apparent correlative of the Brassfield Formation is the Brassfield Limestone, a cherty limestone, generally 20 feet thick. As noted above in the descriptions of the facies of the Ordovician strata in central Tennessee, the Nashville dome had a strong influence on depositional patterns. Farther to the west, however, tectonic influence on depositional patterns became more complex during Silurian and Devonian Periods when several other tectonic elements of the craton became involved (see figure 10.7). One feature exerting influence during this time was the southeastern extension of the Ozark Dome, an uplifted area located mostly in what is presently Arkansas. The Ozark Dome was aligned with the southwestern end of the Nashville Dome by a structurally low region called the Clifton saddle (mainly in what are now Hardin, Decatur, Wayne, and Perry Counties in Tennessee). To the northwest was the Illinois Basin and to the west the Reelfoot (New Madrid) rift. The ups and down of these tectonic elements exerted profound control on depositional and erosional patterns in this region for nearly 100 million years. Isopach maps constructed by C. W. Wilson Jr. (1949) of Ordovician, Silurian, and Devonian formations in central and west central Tennessee show thickening and thinning of units, scoured erosional channels filled with siltstone or sandstone, and in some cases the total absence of formations, all characteristics that can be interpreted as fluctuations of sea level (transgressing/regressing seas and erosion) possibly attributed to ups and downs of these tectonic elements.

Because of the complex array of facies of the Silurian rocks in Middle and West Tennessee, geologists over the years have recognized a number of different subdivisions. In most cases, the subdivisions have resulted from alternating units of limestone and shale. Three stratigraphic units have been recognized above the late Silurian Brassfield Limestone, the Wayne Group, the Brownsport Group, and the Decatur Limestone. Units within the Wayne Group, from oldest to youngest, include the Osgood Formation (mostly greenish-reddish shale, 0 to 15 feet thick), the Laurel Limestone (mostly olive-gray limestone, 0 to 30 feet thick), Waldren Shale (greenish gray fossiliferous shale, 0 to 5 feet thick), Lego Limestone (mostly olive-gray limestone, 0 to 30 feet thick), and Dixon Formation (green and reddish shaly limestone, 0 to 20 feet thick). The Brownsport Group above the Wayne Group consists of the Beech River Formation (shale with thin beds of limestone, 0 to 60 feet thick), the Bob Limestone (thick-bedded limestone, 0 to 25 feet thick) and the Lobelville Formation (shale with beds of limestone, 0 to 30 feet thick). The Decatur Limestone above the Brownsport is a thick-bedded, medium- to coarse-grained limestone (0 to 70 feet thick). According to Gibson (1990) the Decatur in the Western Valley area may have extended farther to the east, but was mainly deposited on a shelf that was bounded on the west by a declivity formed by the Reelfoot rift (see figure 10.7). If the Decatur at one time extended farther to the east, it has subsequently been removed from the record by erosion.

Devonian

By early Devonian time, the epeiric seas appear to have regressed from most of Tennessee, leaving the region subject to extensive erosion; hence, a rock record is scarce. In East Tennessee places can be observed where strata as old as middle Ordovician were channeled by erosion before late Devonian sediments were deposited disconformably upon them. Regarding deposition in West Tennessee, especially in the Western Valley sections on the shelf lying between the Nashville Dome and the Reelfoot rift, it appears that the region remained submerged and that sedimentation continued much in the fashion of the underlying Silurian rocks. Tectonic elements of the region continued to control sediment deposition of the strata above the Silurian Decatur Limestone. As the sea regressed westward (based on the present-day geographic orientation), a mixed carbonate-siliciclastic depositional environment, resulting from carbonate **bioclastic** mud (finely broken shells) being mixed with terrigenous silt and mud derived from nearby uplifted areas, formed where sea currents circulated along the Reelfoot rift and the nearby Illinois Basin.

The fossiliferous Ross Formation (0 to 75 feet thick), above the Decatur Limestone, consists of facies representing several different environments. Each of these facies is a subdivision of the formation and thus considered to be members of the Ross. At the base are coeval members, the Rockhouse Limestone member and the Rockhouse Shale member, and above the Rockhouse is the very fossiliferous Birdsong Shale member.

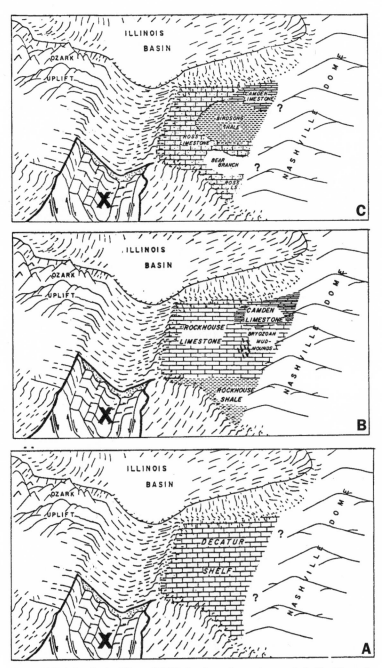

Figure 10.7. A drawing by Gibson (1988) showing paleoenvironments in West Tennessee as they might have appeared during late Silurian time (A. Decatur Limestone) and early Devonian time (B. Rockhouse time and C. Birdsong Shale time). The Nashville Dome was possibly an uplifted area forming a barrier isolating the western part of the state from the east. The cutaway (X) shows inferred structure beneath the Reelfoot rift (the structure responsible for the great 1811–1812 earthquakes).

Either the Harriman or Camden, two formations that are very similar and only separable on the basis of paleontology, overlies the Ross Formation. Lithologically the Harriman and the Camden are both **novaculitic** (very dense and even-textured cryptocrystalline) chert and tripolitic clay (rock composed of shells of diatoms and radiolaria or disintegrated chert) with minor siliceous limestone (variable thicknesses ranging from 0 to 100 feet). Gibson (1990) postulated that the influx of silica into the depositional environment is from a volcanic source to the east. This explanation is feasible in that collisions of terranes (some volcanic islands) with proto–North America are believed to have been continuous through the Devonian Period beginning with the Taconic orogeny. The shallow sea in the Western Valley region regressed following deposition of the Camden, and the land was eroded before the next advance of the sea. The youngest Tippecanoe unit, the Pegram Formation, a thick-bedded, gray limestone and sandstone (0 to 30 feet thick), was deposited on the eroded surface of the Camden and older units. Following deposition of the Pegram and its correlative units, proto–North America, including Tennessee, became land and was deeply eroded. As previously mentioned, in places erosion cut down into bedrock as deep as middle Ordovician strata forming the erosional surface forms of an unconformity that marks the close of the Tippecanoe Sequence.

11

Kaskaskia Sequence

> How trivial a thing a rotten shell may appear to some,
> yet these monuments of nature are more certain tokens
> of antiquity than coins or medals.
>
> —Robert Hooke, as quoted by Charles Lyell (1832)

The Devonian/Mississippian Chattanooga Shale

Blanketing the deeply eroded surface of previously deposited Devonian sediments is a ubiquitous, radioactive black shale known in Tennessee as the Chattanooga Shale (see figure 11.1). It is correlated with other black shales of varying thicknesses across many regions of North America where it has been given a variety of names. Paleontological evidence suggests that the Chattanooga is possibly both late Devonian and early Mississippian in age—being older in the east and becoming progressively younger to the west as the Chattanooga sea transgressed from east to west. The black color is the distinguishing feature of the Chattanooga—such that it is often referred to as the Chattanooga black shale. The black color makes the unit easily recognized and a convenient marker for the time boundary between the Tippecanoe and Kaskaskia Sequences (Devonian and Mississippian Periods).

In sedimentary rocks a black color is generally indicative of either high organic content and/or **euxinic** conditions (poor circulation and/or low oxygen). The fact that this depositional environment is so widespread over the eastern portion of North America continues to perplex geologists. **Fossil conodonts** (microscopic worm jaws), *Tasmanites* (plant spores), **linguloid brachiopods** (lamp shells), and possible wood fragments represent the meager preserved biota in the shale; however, evidence is lacking for a cause of restricted flow. Some geologists have speculated it was deposited in stagnant basins, perhaps thousands of feet deep, and others suggest shallow water over vast mudflats. Regardless of how deep the sea might have been, the rich black color and the presence of **iron disulfide** (the minerals pyrite/marcasite) strongly suggest decomposition of organic

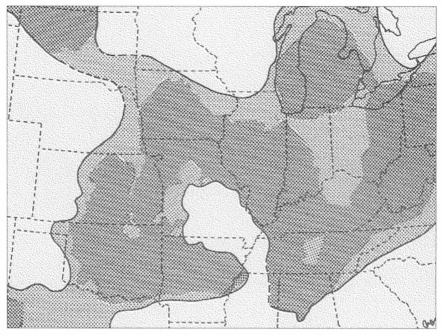

Figure 11.1. Extent of the Chattanooga Shale and its correlatives in portions of eastern and mid-continental United States. Darker shaded area represents present extent, and the lighter shaded area represents probable former extent. (Conant, Swanson 1961.)

matter to produce hydrogen sulfide in an oxygen-deprived environment. Modern analogs for both environments exist today in various parts of the world, e.g., the Black Sea. Even though the Chattanooga Shale once blanketed a large portion of Tennessee, because of the deformation during the Alleghanian orogeny, exposures of the Chattanooga in eastern Tennessee are limited almost entirely to the base of the eastern escarpment of the Cumberland Plateau, where faulting has brought the unit to the surface. And with the exception of scattered exposures surrounding the Nashville basin on the Highland rim, erosion has removed its record over most of western Tennessee.

The thickness of the Chattanooga varies geographically across Tennessee, ranging from 0 up to 900 feet. Typically it is 20 to 30 feet thick. Thicker sections of Chattanooga and its equivalents lie to the northeast, where in eastern Virginia and Pennsylvania it is thousands of feet thick. One exception in Tennessee, where the Chattanooga thickness exceeds the typical tens of feet in thickness, is the Flynn Creek meteorite impact crater in Jackson County, Tennessee (see chapter 15), where at least 200 feet of the Chattanooga sediment accumulated in the depression created by the impact. Using the principle of relative dating determines that the impact is younger than the age of the youngest rock impacted—Ordovician—and older than the oldest rock not impacted—Devonian Chattanooga Shale.

The rock is radioactive and has been studied over the years as a potential source of uranium. At the present, however, it is not considered as an ore—despite large reserves, the grade of the ore is very low. The Chattanooga, however, does have economic value as a source rock for hydrocarbons. Oil and gas trapped in reservoir rocks on the Cumberland Plateau are believed to be derived from the underlying Chattanooga Shale (see chapter 16).

Mississippian

The Mississippian rocks represent the last great incursion of a seaway across Tennessee. As the waters became less muddy, the Chattanooga Shale graded upward into a thin (maybe a foot thick) robin's-egg blue shale containing nodules of phosphate. This is the Maury Shale, which is typically considered as the basal part of the Fort Payne chert (200 to 300 feet thick). The Fort Payne is a calcareous and dolomitic silicastone containing bedded chert, cherty limestone, dolomite, and shale, and locally, especially in the subsurface of the Cumberland Plateau, **bioherms** (reeflike structures) of pelmatazoan that form reservoirs for hydrocarbons. The high silica content of the Fort Payne provides erosional resistance much like the chert in Cambrian-Ordovician Knox Group formations, and, where the formation is tilted, it typically erodes to form ridges called hogbacks. The hogback ridges in the Valley and Ridge province of East Tennessee are so typical that the formation can readily be identified by its topographic expression. Perhaps the best example of the Fort Payne's resistance to erosion is where it is gently dipping or relatively flat-lying and forms the resistant cap for the Highland Rim surrounding the Nashville basin.

The early Kaskaskian shallow sea that transgressed across Tennessee created environments conducive to rich biota of coral, pelmatazoa, gastropods, brachiopods, bryozoa, and many other forms of life and can be compared in many ways to the great inland sea of Ordovician time. The Mississippian Period is often referred to as the "age of crinoids" (see figure 11.2). Limestone is the dominant Mississippian rock type, but a few sandstones and shales representing periodic influxes of siliciclastic material occur throughout the section. Figure 11.3 is a generalized stratigraphic section of Mississippian formations in Tennessee. The thickness of the entire section ranges from about 700 feet in Middle Tennessee up to as much as 5000 feet in East Tennessee. Although Mississippian rocks once extended across what is now Tennessee, they have essentially been removed by erosion. The best representative sections of Mississippian strata lie protected

Figure 11.2. The Mississippian Period is often called the "age of crinoids." Crinoids, or sea lilies as they are often called, were dominant, and their fragments (often called Indian beads) often are the major components of limestone. Because echinoids other than crinoids had similar stems, beadlike pieces of broken stem are collectively called **pelmatazoa**. Bryozoa, seen here attached to the seafloor, were also abundant during this period. Fish, as seen in the picture, became diverse during this time. (Exhibit by Chase Studios, Inc. Courtesy of the McClung Museum, the University of Tennessee.)

beneath the Cumberland Plateau, cropping out along its eastern and western escarpments where streams have incised their valleys by headward erosion into the plateau.

Eastern proto–North America was tectonically active as a convergent plate boundary during Devonian time. In the present-day New England region, the convergence caused the Acadian orogeny forming the ancient Acadian Mountains. And by late Mississippian time the effects of collisions with the continent of Africa migrated southward, and the ancestral Blue Ridge Mountains, long ago worn down following the Taconic orogeny, were uplifted again. As new mountains began to emerge, sediment derived from them gradually began to muddy the water, first in the Appalachian Basin and then the shallow seas to the west.

	Cumberland Plateau Highland Rim & Western Valley Ridge	Eastern & Northeastern Valley & Ridge Tennessee
Upper Mississippian	Pennington Formation	Pennington Formation
	Bangor Limestone	Newman Limestone— Greasy Cove Formation
	Hartselle Formation	
	Monteagle Limestone	
	St. Louis Limestone	
	Warsaw Limestone	
Lower Mississippian	Fort Payne Formation	Grainger Formation
	Maury Shale	

Figure 11.3. Formations that typically make up the Mississippian System of Tennessee.

Thus shales and sandstones gradually replaced the life-bearing limestone that was deposited in the once shallow, clear seas.

Eastern North America continued to rise as Africa was subducted beneath it, sending rivers laden with sediment out into Mississippian seaway as floodplains and deltas began to form along the margin of the sea. As the Appalachians continued to rise, deposition of the late Mississippian Pennington Formation continued the environmental transition from sedimentation in a marine environment to that of a terrestrial environment. **Paleosols** (ancient soils) and karst features identified in the Pennington Formation (Caudill and others 1996) provide evidence that **supratidal** (above tides) land surfaces existed periodically during the gradual transition from marine to terrestrial. The Pennington grades upward into Pennsylvanian-aged rocks that were forming as a clastic wedge prograding from east to west and marking the close of the Kaskaskian Sequence and the beginning of the Absaroka Sequence.

The Maury Shale, about a foot thick, is bluish-green shale that contains brown phosphate nodules. Fort Payne Formation, 100 to 275 feet thick, is calcareous and dolomitic silicastone composed of bedded chert, cherty limestone, and shale. The Fort Payne is one of those formations whose topographic expression is notable. When traveling west toward the Nashville Basin a broad relatively flat but rolling land surface is encountered that must be crossed before dropping down into the basin. This broad area is the Highland Rim, which is underlain by the Fort Payne. Superposed on the Fort Payne is the Warsaw Limestone, a 100- to

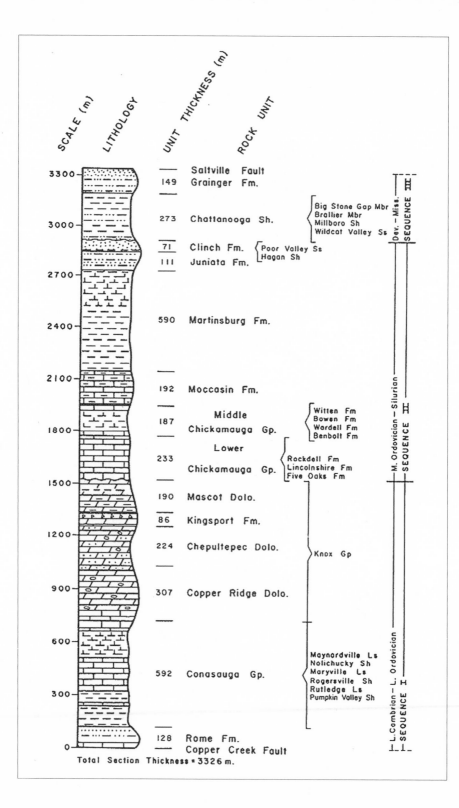

SCALE (m) — LITHOLOGY — UNIT THICKNESS (m) — ROCK UNIT

Scale (m)	Unit Thickness (m)	Rock Unit	Members/Formations	Age	Sequence
3300		Saltville Fault			
	149	Grainger Fm.			SEQUENCE III
3000	273	Chattanooga Sh.	Big Stone Gap Mbr / Brallier Mbr / Millboro Sh / Wildcat Valley Ss	Dev.–Miss.	
	71	Clinch Fm.	Poor Valley Ss / Hagan Sh		
	111	Juniata Fm.			
2700					
2400	590	Martinsburg Fm.		M. Ordovician–Silurian	SEQUENCE II
2100	192	Moccasin Fm.			
1800	187	Middle Chickamauga Gp.	Witten Fm / Bowen Fm / Wardell Fm / Benbolt Fm		
	233	Lower Chickamauga Gp.	Rockdell Fm / Lincolnshire Fm / Five Oaks Fm		
1500	190	Mascot Dolo.			
1200	86	Kingsport Fm.			
	224	Chepultepec Dolo.	Knox Gp	L. Cambrian–L. Ordovician	
900	307	Copper Ridge Dolo.			SEQUENCE I
600					
	592	Conasauga Gp.	Maynardville Ls / Nolichucky Sh / Maryville Ls / Rogersville Sh / Rutledge Ls / Pumpkin Valley Sh		
300					
0	128	Rome Fm.			
		Copper Creek Fault			

Total Section Thickness = 3326 m.

130-foot-thick, cross-bedded, medium- to coarse-grained, gray, sandy limestone. Overlying the Warsaw is the St. Louis Limestone, a fine-grained, brownish-gray limestone, dolomitic and cherty and 80 to 160 feet thick. The Monteagle Limestone, 180 to 300 feet thick, is a light-gray oölitic limestone containing abundant pelmatazoan fragments. Overlying the Monteagle and creating a break in the limestone sequence is the Hartselle Formation, a thin-bedded, fine-grained, calcareous sandstone interbedded with gray shale and oölitic and coarse-grained limestone beds. Although only up to 80 feet in thickness where present, the Hartselle generally forms a topographic bench on escarpments of the Cumberland Plateau.

The Bangor Limestone, a 70 to 400 feet thick, dark-brownish-gray thick-bedded limestone, was the last carbonate to form as the shallow Mississippian sea was gradually being filled with clastic sediment. The limestone of the Bangor grades upward into the previously described shaly Pennington Formation, 150 to 400 feet thick. The reddish and greenish shale, as well as the siltstone, dolomite, dark gray limestone, and thin-bedded sandstone of the Pennington, reflects the gradual transition from a marine to a terrestrial environment. In eastern and northeastern Tennessee, a thick (500 to 1000 feet), gray to green shale, the Grainger Formation is probably equivalent to the Fort Payne Formation. The Grainger represents distal sediment, a delta prograding from Alleghanian uplifting in central Virginia. Likewise, east of the plateau and in the vicinity of Clinch Mountain, where the Mississippian limestone is subdivided into units such as the Bangor or Monteagle, it is generally mapped as the Newman Limestone. Where the Newman is absent, the Grainger Formation is overlain by the Greasy Cove Formation, a gray, argillaceous limestone, calcareous shale, and siltstone that ranges up to 400 feet in thickness.

The "Thorn Hill Section," perhaps the most complete slice of geologic history in Tennessee, is exposed along U.S. Highway 25-E where it crosses Clinch Mountain in Grainger County. The section embraces rocks from the Sauk Sequence through the Kaskaskian Sequence. Figure 11.4 represents the nearly continuous exposures along the highway. For a complete description of this section the reader is directed to Byerly and others (1986) and Walker and others (1985).

Figure 11.4., *opposite.* What is perhaps the most complete stratigraphic section in Tennessee, the so-called Thorn Hill Section, is represented in this drawing. Exposed along U.S. Highway 25-E in northwest Grainger County from near Thorn Hill and across Clinch Mountain, the layers of rocks in this section range from the Cambrian System of the Sauk Sequence to the Mississippian System of the Kaskaskian Sequence.

12

Absaroka Sequence

> Those who dwell, as scientists or laymen, among the
> beauties and mysteries of the earth are never alone or
> weary of life.
>
> —Rachel Carson, *The Sense of Wonder* (1956)

Pennsylvanian

Even though there are places on the North American craton where a **disconformity** (profound erosional surface) separates the Pennsylvanian System (Absaroka) from the Mississippian System (Kaskaskia) over most of North America, the boundary is equivocal. Such is the case in Tennessee, where late Kaskaskian rocks grade upward into early Absarokan strata with no clear boundary. For this reason many geologists prefer lumping the two systems into the "Carboniferous" Period, the initial term coined in Europe to refer to this interval of Earth history during which coal was formed. Early geologic mapping in the upper Mississippi River valley interpreted marine limestone unconformably overlain by the coal-bearing rocks formed in a terrestrial environment to represent two separate periods—the marine limestone was called Mississippian, and the coal-bearing rocks were called the Pennsylvanian. Nevertheless, in keeping with the sequence stratigraphy scheme chosen to simplify discussion of the rock record in this book (see chapter 7), the sequences Kaskaskian and Absaroka Systems will be retained, and the Pennsylvanian System in Tennessee will represent the Absaroka Sequence. However, a major problem in reckoning with the boundary between the two systems/sequences is that the time boundary occurs somewhere within a transition from deposition on a marine carbonate shelf (Mississippian limestone) through a sequence of marine/terrestrial red-green shales, sandstones, and minor carbonates (Pennington Shale), to coal-bearing shales, siltstones, sandstones, conglomerates, and minor carbonates (Pennsylvanian). There is no widespread physical break between the two sequences.

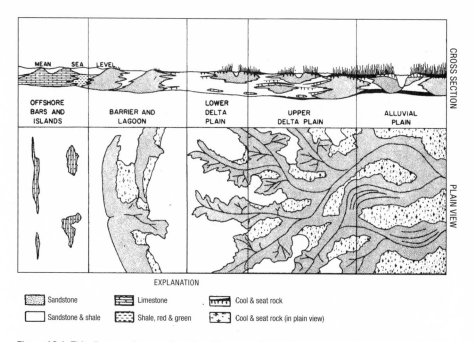

EXPLANATION

Sandstone   Limestone   Cool & seat rock

Sandstone & shale   Shale, red & green   Cool & seat rock (in plain view)

Figure 12.1. This diagram shows a plan view of a prograding delta such as may be observed in a modern deltaic setting such as the Mississippi Delta. The cross-section shows how different types of sediment accumulate contemporaneously in different settings in the model to form sedimentary facies. (Modified from Ferm JC 1974.)

Although probably once extending across Tennessee, most of the Absaroka record has been removed by erosion, with the exception being the bedrock of the Cumberland Plateau. At least two examples verify a more widespread occurrence of Pennsylvanian rocks across Tennessee: Short Mountain, in Cannon County, Tennessee, on the western edge of the eastern Highland Rim and about 20 miles west of the nearest outcrop of Pennsylvanian rocks on the plateau to the east; and Grindstone Mountain, in Hamilton County, Tennessee, in the middle of the Valley and Ridge province and about 17 miles west of the nearest Pennsylvanian outcrop on the eastern escarpment of the plateau.

Topography and relative ages of rocks have warranted dividing the Cumberland Plateau where Pennsylvanian rocks are dominant into a southern section and a northern section (sometimes called the Cumberland Mountains). The Emory River (Roane and Morgan Counties) follows the trend of a northwest-southeast strike-slip fault that separates the two sections; however, highway I-40 serves as an arbitrary boundary separating the relatively flat upland bisected by Sequatchie Valley at the southern end from the more mountainous northern end. Although the northern section is more mountainous, it lies within a structural

Figure 12.2. Cross-bedded Pennsylvanian sandstone in a fluvial (stream) channel eroded into shale that was once mud of an ancient lagoonal environment.

low referred to as the Wartburg Basin. This structural condition in the northern section presents rocks at the land surface younger (middle Pennsylvanian) than those exposed on the southern plateau (early Pennsylvanian).

Early geologists who mapped Pennsylvanian strata tried, with great difficulty, to correlate the coal beds, sandstones, and shales across the Tennessee region, as well as with Pennsylvanian rocks in other regions of North America. For the most part, these early workers regarded the strata as though they were like a stack of pancakes, laterally equivalent everywhere. The problem inherent here is that the concept of sedimentary facies—recognizing that rock units with varying physical properties grade laterally into one another, which reflects lateral gradation of one ancient depositional environment into another—was not well known by many of the early geologists. For example, consider a barrier beach such as exists along the coast of Georgia or the Carolinas (most vacation beaches in those states are barrier beaches). Behind the barrier beaches lie salt marshes or lagoons and also tidal inlets that likely cut through the barriers. Each of those environments simultaneously receives a different type of sediment, and those sediments grade laterally one into the other with the result eventually being that

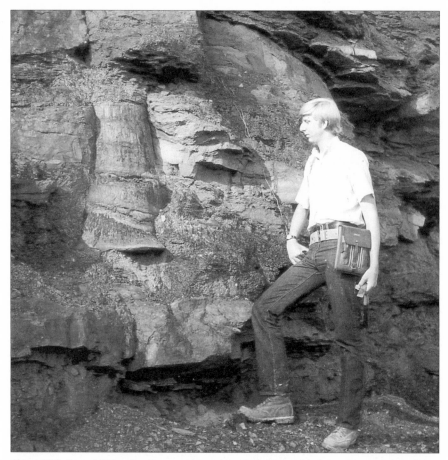

Figure 12.3. An upright tree fossil exposed in a strip mine's high wall in the northern section of the Cumberland Plateau. The Pennsylvanian strata were deposited in an upper delta plain paleoenvironment where the trees were likely growing on an ancient floodplain.

when they are buried and lithified, a rock unit will be formed that may be sandstone in one part and shale in another (see chapter 7). Pennsylvanian rocks in Tennessee best fit barrier-back, barrier-lagoon, and deltaic-fluvial models of deposition. This means a scenario where rivers with floodplains (the fluvial aspect of the model) carried sediment presumably in a southwesterly direction across what is now Tennessee to shorelines complete with beaches, sandbars, barrier beaches, tidal marshes, and deltas that prograded or migrated westerly. All these paleoenvironments are subject to relatively rapid changes over geologic time, resulting in complex facies relationships. You may recall that complex facies are associated with similar quickly changing conditions on the Ordovician carbonate shelf paleoenvironments.

Figure 12.4. Roots of a Pennsylvanian tree preserved in rocks formed in an upper delta plain paleoenvironment on the Cumberland Plateau. The fossils of the roots are known as stigmaria.

The Pennsylvanian strata of each section of the plateau reveal different sets of paleoenvironments as sediments derived from northern and eastern uplifted areas prograded across the interior of proto–North America. The older sediments of the southern Cumberland Plateau were primarily deposited as barrier beaches, tidal lagoons, and deltas at the leading edge of the progradation, whereas the northern Cumberland Plateau sediments accumulated on upper delta plains with flood plains, natural levees, levee washovers, and oxbows. Figure 12.1 is a diagram showing environments related to a pro-grading delta. Figure 12.2, a photograph of a conspicuous outcrop on I-40 near mile marker 299, shows cross-bedded sandstone deposited in a fluvial (stream) channel that eroded through shale that was once mud in a lagoonal environment. As progradation proceeded across the continent, the ocean environment was progressively replaced by a terrestrial environment. The Pennsylvanian rocks exposed on the southern plateau were formed in a transitional environment, partially marine and partially terrestrial, as deltas were prograding into a shallow sea, and the younger Pennsylvanian rocks of the northern plateau were formed in terrestrial environments. This environmental difference is reflected in the fossil vegetation. Fossil trees in strata of the southern plateau appear to have accumulated mostly as driftwood, whereas many of the tree fossils in the northern section are rooted and in upright growth position (see figs. 12.3 and 12.4).

Middle Pennsylvanian	Cross Mountain Formation	
	Vowell Mountain Formation	
	Redoak Mountain Formation	
	Graves Gap Formation	
Lower Pennsylvanian	Indian Bluff Formation	
	Slatestone Formation	
	Crooked Fork Group	Wartburg Sandstone
		Glenmary Shale
		Coalfield Sandstone
		Burnt Mill Shale
		Crossville Sandstone
		Dorton Shale
	Crab Orchard Mountain Group	Rockcastle Conglomerate
		Vandever Formation
		Newton Sandstone
		Whitwell Shale
		Sewanee Conglomerate
	Gizzard Group	Signal Point Shale
		Warren Point Sandstone
		Raccoon Mountain Shale

Figure 12.5. Pennsylvanian rock units in Tennessee. The lower Pennsylvanian Gizzard and Crab Orchard Mountain groups underlie and are the exposed rocks of the southern Cumberland Plateau and the Crooked Fork Group. Younger units of middle Pennsylvanian age are exposed in the northern Cumberland Plateau.

Pennsylvanian stratigraphic nomenclature in Tennessee has apparently not been resolved. The earliest work described the stratigraphic section in discreet formational units (the pancake analogy). Later these formations were assembled into groups; however, this did not resolve the problem. The Tennessee State Geologic Map (Hardeman 1966) describes part of the section (mostly the older rocks in the south) as groups and treats the middle Pennsylvanian as ungrouped formations. The problem results from facies—rock types are environmentally controlled, not time controlled. For example, Warren Point Sandstone at Point Park on Lookout Mountain had once been correlated with the younger Sewanee Conglomerate on Walden Ridge, because both have similar physical attributes. Figure 12.5 is a generalized stratigraphic column of Pennsylvanian rocks in Tennessee according to the Geologic Map of Tennessee (Hardeman 1966).

13

Alleghanian Orogeny

Beautiful is what we see. More beautiful is what we understand. Most beautiful is what we do not comprehend.

—Nicolaus Steno, 1673

Tectonic activity leading up to the grand finale mountain-building episode in the Appalachians—the Alleghanian orogeny—actually began about 470 Ma during the Taconic orogeny in the Ordovician period and continued through the Devonian with the Acadian orogeny when the Iapetus Ocean began to close, and various continental plates such as Gondwana (Africa, Australia, India, and South America) and Baltica (northern Europe not including Ireland and Scotland) started migrating westward toward proto–North America (see figure 10.1). Scattered between the proto–North America plate and the large plates Gondwana and Baltica were a number of smaller pieces of exotic real estate called terranes or microcontinents. These too were on course for collision with proto–North America—colliding throughout late Paleozoic time. Some of these smaller terranes were actually volcanic islands formed when parts of the Iapetus seafloor were subducted and heated sufficiently to form magma that rose and erupted to form volcanic islands.

Neither Gondwana nor Baltica collided with proto–North America during the early stages of tectonic activity; however, some volcanic islands and terranes were sutured onto proto–North America to form part of what is now the Piedmont physiographic province east of the Blue Ridge province of Tennessee and North Carolina. Deformation related to these collisions was sufficient to form the Taconic Mountains, which may have stretched from New England to Alabama; however, they have subsequently been eroded away, and now all that records their existence are the rocks derived from sediment eroded from them (see chapter 10).

The Iapetus Ocean continued to close and then, some time during the Devonian period about 380 Ma (Acadian orogeny), Baltica collided with the northeastern part of proto–North America to form another mountain chain called the Acadian Mountains. Not only did Baltica collide with proto–North America, but also several microcontinents or terranes collided as well—namely the Avalon terrane, also known as Carolina terrane, in the southern Appalachians. Evidence of this orogeny in Tennessee is similar to that of the Taconic orogeny—it is mainly preserved in the sedimentary record as sediment eroded from the mountains. Sea-level fluctuations and erosion of the Acadian Mountains during the orogeny provided sediment for late Tippecanoe (Devonian) and early Kaskaskian (Mississippian) rocks in the Appalachian Basin. Further evidence of the orogeny is observable in the metamorphic rocks and igneous plutons in the Blue Ridge province.

In a broad sense, then, the Alleghanian orogeny began with the Taconic orogeny, continued with the Acadian orogeny, and climaxed when Gondwana collided with proto–North America some 250 to 300 Ma. All during the time of the Taconic and Acadian Orogenies, Gondwana continued to glide closer and closer to proto–North America until it ultimately collided and formed the supercontinent Pangea. In Tennessee the effects of the orogeny are most profound in the Blue Ridge and the Valley and Ridge provinces. In the Blue Ridge, rocks deformed during the Taconic orogeny were deformed again during the Acadian event, and then these "deformed deformed" rocks were deformed at least a third time during the Alleghanian climactic event. The rocks of the Blue Ridge are metamorphosed, and the rocks in the Valley and Ridge, though not metamorphosed, were intensely folded and faulted during the Alleghanian orogeny.

Keep in mind, tectonic plates move at excruciatingly slow rates. The ancient Appalachian Mountains, perhaps equal in heights to those of the Himalayan Mountains, were built at such a rate throughout the Pennsylvanian and Permian Periods (Absaroka time) that erosion attacked the uplifted mountains as they were being born and that process continues today. Billions and billions of cubic yards of sediment removed from the mountains can be found today as far away as the Mississippi Delta. The collision pushed what are now the Smokies and the Valley and Ridge possibly a 100 miles or more westward from their original geographic positions along a décollement (sole fault), piling layers of rock upon one another in an **imbricate** (overlapping) fashion, like shingles on a roof.

The Blue Ridge Province

In Tennessee the effects of the Alleghanian orogeny are most obvious in the Blue Ridge physiographic province. Here the rocks have been intensely folded, faulted, and metamorphosed. Metamorphic zones showing the intensity of metamorphism as mapped on the Geologic Map of Tennessee (Hardeman 1966) are, beginning from west (low grade) to east (high grade), chlorite, biotite, and garnet

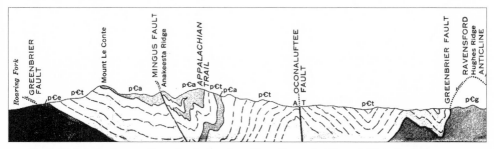

Figure 13.1. A structure section extending northeast from Ravensford through Mt. LeConte to Roaring Fork in Great Smoky Mountains National Park. At least two episodes of deformation can be implied from this section. Note that the Greenbrier fault is folded. The leading hypothesis is that the folding occurred after the faulting took place. This would indicate two episodes of deformation. Adding to that fact is that this folded fault was later transported to the west on the hanging wall to the Great Smoky fault. Thus three episodes of deformation are indicated: pЄe = Elmont Sandstone; pЄt = Thunderhead Sandstone; pЄa = Anakeesta Formation; and pЄg = Grenvillian rocks. (King, Neuman, Hadley 1968.)

(see chapter 3). Since Africa (Gondwana) bumped us from the east, this is what one would expect.

The Grenville basement rocks exposed in northeastern Tennessee, having endured the most episodes of **dynamothermal** (pressure and heat) metamorphism associated with the orogenies from Grenville through Alleghanian, exhibit the highest grade metamorphism observed in Tennessee. As described in chapter 8, these rocks are dominantly crystalline granite gneiss.

The rocks of the Ocoee Supergroup that form the bedrock of the Smoky Mountains are regarded as metasedimentary rocks, meaning that, although the shales have been metamorphosed to slates or phyllites and the coarser clastic sediments like sandstone, graywacke, or conglomerate, the rocks have retained most of their sedimentary structures, such as cross-bedding and graded bedding. Hence, these rocks are typically called metasandstone, metagraywacke, and metaconglomerate. The same can be said of the Sauk Sequence rocks in the Mount Rogers area. The rocks of the Chilhowee Group tend to exhibit lower grade metamorphism than those of the Ocoee Supergroup, but can still be considered metasedimentary. A notable example is the metamorphic **orthoquartzite** (silica-cemented sand) deposited as clean beach sand on the passive continental margin during Chilhowee time (see figure 3.5).

The most conspicuous aspect of rock deformation in the Blue Ridge is the presence of multiple partings or fractures in the bedrock. These partings occur in the form of original bedding (that started out as horizontal), foliation that includes rock cleavage (as in slaty), joints, and faults. The cross-cutting relationships of these elements attest to a minimum of at least three episodes of deformation. Perhaps the most straightforward example of this is illustrated in the Great Smoky Mountains by the relationship of the Greenbriar fault (perhaps the oldest in the Smokies) and the Great Smoky fault. Sequentially, the Greenbriar thrust fault underlying Mount LeConte in the Smoky Mountains was emplaced first; it was then folded along with the hanging wall and footwall rocks, and finally the folded fault was transported westward along the Great Smoky thrust fault during the Alleghanian orogeny (see figure 13.1).

An interesting phenomenon associated with the folding and faulting in the Blue Ridge is the development of **duplex faults**. A duplex is a system of imbricate thrusts that have split off from a main fault below and merge with a fault above to form a stack of thrust-bounded rocks (see figure 13.2). A number of duplexes formed in the younger Cambrian/Ordovician carbonate rocks in the footwall of the Great Smoky fault, forcing the fault plane of the Great Smoky fault

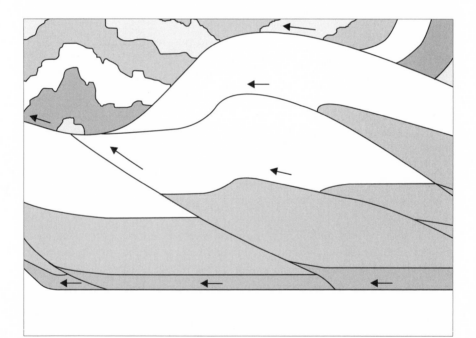

Figure 13.2. An example of duplexing faults. As the thrust faults ramp upward, each successive fault is forced higher in a piggyback fashion. Such duplexing caused a major fault such as the Great Smoky fault to glide above the duplexes (i.e., the upper fault in the diagram). Once erosion breached the upper fault, erosion continued downward into more easily eroded rocks such as limestone (light colored layers in the diagram) below the fault plane, creating structural features known as windows (fensters) such as Cades Cove and Tuckaleechee Cove. Also, note in the diagram that the lower fault plane represents the décollement (sole fault), and the darker layer represents the Rome Formation that is repeated in thrust faults splaying off the decollement toward the left-hand side of the diagram.

to bow upward above the duplexes. Because the duplexed rocks were elevated and dominantly carbonates, they became prone to erosion by dissolution, and the Great Smoky fault, being relatively flat, was soon eroded through, exposing the younger duplexed footwall rocks below. Structurally these areas are now referred to as **windows** or **fensters,** where the older rock of a hanging wall is eroded through and now surrounds the exposed younger rocks of the footwall. We recognize these windows today as coves, the most famous of which is Cades Cove in Great Smoky Mountains National Park. Other windows of note in the Blue Ridge of Tennessee include Wear Cove, Tuckaleechee Cove (see figure 13.3), and Miller Cove.

As previously mentioned, the original Appalachian Mountains may have been comparable in stature to the present-day Himalayan Mountains, which are still growing. The Tibetan Plateau and the Himalayan Mountains continue

Figure 13.3. Duplex faults in Tuckaleechee Cove near Townsend, Tennessee. Note the two sets of arched limestone beds of the Ordovician Jonesboro Formation near the upper right-hand corner of the photograph. They are duplexed onto one another. A fault dips to the southeast (right) diagonally across the center of the photograph along a line marked by trees. Precambrian rocks of the hanging wall of the Great Smoky fault form the hump at the very top of the ridge. The hump is bounded by the fault and exists as an outlier or klippe (older rocks bounded by a fault and resting upon younger rocks—the opposite of a window or fenster).

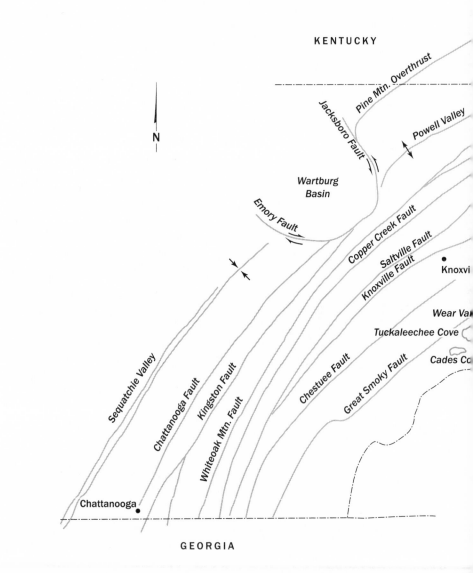

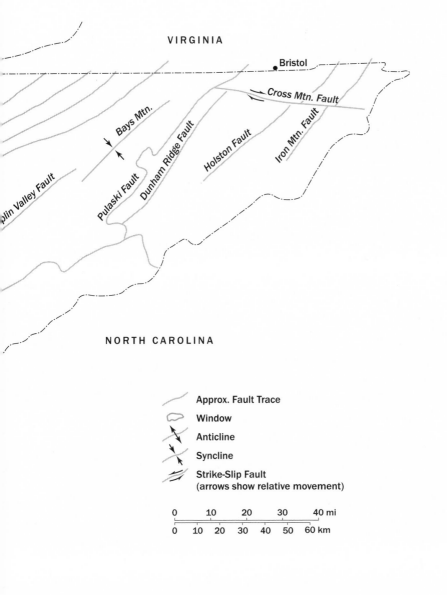

Figure 13.4. A map of East Tennessee depicting the traces of some of the major thrust faults and folds associated with Alleghanian orogeny. (Modified from Rodgers J, 1953.)

VIRGINIA

Bristol

Cross Mtn. Fault

Bays Mtn.

Dunham Ridge Fault

Holston Fault

Iron Mtn. Fault

plin Valley Fault

Pulaski Fault

NORTH CAROLINA

Approx. Fault Trace

Window

Anticline

Syncline

Strike-Slip Fault
(arrows show relative movement)

0   10   20   30   40 mi

0   10   20   30   40   50   60 km

to rise because they are buoyed up by the continental crust of the Indian Plate as it continues to be subducted beneath the Asian plate. This buoying happens because the continental crust of the Indian plate is less dense and more buoyant than oceanic crust. Likewise it is conceivable that continental crust from Gondwana (Africa) was subducted beneath proto–North America, buoying the Appalachians upward during the Alleghanian orogeny. Although the Appalachian Mountains we see today have their beauty, they are but the remnants of a once-spectacular mountain range that has been dissected by more than 200 million years of erosion.

The Valley and Ridge Province

The Valley and Ridge province is just that, alternating ridges and valleys formed through the differential erosion of early and middle Paleozoic carbonate and noncarbonate rocks that have been stacked in imbricate fashion like shingles on a roof that are arranged in elongate NE-SW trending belts by Alleghanian deformation. Figure 13.4 is a map of East Tennessee produced by Rodgers (1953) showing the many thrust faults related to the Alleghanian orogeny. The Rome Formation, which serves as the major detachment fault or **décollement** along which everything was transported westward, broke upward at a number of locations to form splay faults across what forms the Tennessee valley region. Each splay brings the Rome Formation to the surface, along with its overlying units, forming the hanging walls of each fault. In this way stratigraphic sections are reproduced over and over, usually with the Rome at the base of each emerging fault (see figs. 1.7 and 13.2). With the exception of a few anticlines and synclines, the imbricate faults have resulted in rocks striking northeast and **inclined** (dipping) to the southeast. Lookout Mountain in Chattanooga and Bays Mountain in upper East Tennessee are two examples of synclines, and the Powell Valley anticline, extending through Campbell, Union, Claiborne, and Hancock Counties, is a good example of an anticline.

When considering the fact that **fault traces** (the surface exposure of a fault) of many Valley and Ridge province faults extend continuously for more than 100 miles, the masses of rock transported along individual faults were huge. The faults farthest from the plate collision and nearest to the Cumberland Plateau show evidence that rupturing tended to ramp upward through strata to glide zones above the Rome Formation—in formations such as shales in the Conasauga Group or the Chattanooga Shale. Farthest to the west, beneath the Cumberland Plateau, deformation dwindles and the glide zone lies mainly within coal beds (see figs. 13.5 and 13.6).

Figure 13.5. Décollement within Pennsylvanian rocks of the Cumberland Plateau west of Dunlap, Tennessee. The dark layer that thickens and thins is a cleaved coal bed along which movement had occurred. The student is pointing to a thrust fault that merges upward with the décollement.

Figure 13.6. A close-up of the cleaved coal of the décollement in fig. 13.5. The rock above the coal is the hanging wall and moved westward (to the left) relative to the rocks below the coal (the foot wall).

The trace of one of the major thrust faults in Tennessee, the Chattanooga fault, extends from the vicinity of Chattanooga to the northeast along the base of the Cumberland Plateau (Walden Ridge) to near Lake City. There it turns almost 90 degrees to the northwest to become the Jacksboro fault, which has both thrust and strike-slip components of movement. The Jacksboro fault extends about 20 miles before it merges with the Pine Mountain thrust in Elk Valley. The Pine Mountain fault strikes northeast along the base of Pine Mountain for about 120 miles through Elk Valley and into Kentucky, where it merges with the Russell Fork fault, the trace of which more or less parallels the trace of the Jacksboro fault (see figure 13.4). This rectangular flap that appears to have been pushed farther to the west than other faults is referred to as the Pine Mountain **overthrust** (a very low-angle thrust fault). Structural windows through the overthrust in Kentucky reveal the Chattanooga Shale as the almost horizontal glide zone.

When traveling north on I-75 along the trace of the Jacksboro fault through Bruce Gap, north of Caryville, Tennessee, one can observe to the northeast several spectacular vertical walls of conglomeratic sandstone separated by intervals of trees known as the "Devil's Racetrack" (see figure 1.10). These tabular-shaped slabs, standing vertically on end and aligned with the crest of the northeast-trending ridge known as Cumberland Mountain, were once nearly flat lying or horizontal. The wide intervals of trees separating the sandstone walls resemble

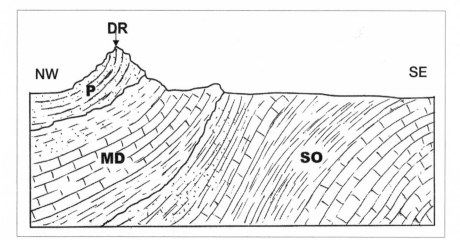

Figure 13.7. Cross-section sketch showing stratigraphy and structure of the Devil's Racetrack (not to scale). P = Pennsylvanian sandstone; MD = Mississippian shales and limestone; SO = Silurian and Ordovician shales and limestone. DR = Devil's Racetrack.

vertical raceways, thus the name. The tree-filled intervals are a result of concentrated weathering and erosion along the vertical planes separating the beds of rock. Water accumulates and moves along the bedding planes causing them to become enlarged through weathering and erosion. The near vertical position of the beds and the resistance sandstone offers to erosion are the probable reasons why Cumberland Mountain exists and the exposure is so spectacular. The beds are the northwestern flank of the Powell Valley anticline (see figure 13.7). The sandstone layers of the Sewanee Formation are nearly vertical, and the bedrock below the rubble on the slopes is mostly shale and limestone—both rock types less resistant to weathering and erosion than sandstone.

The Cumberland Plateau

As previously mentioned, evidence of the Alleghanian orogeny diminishes westward. However, faulting within Pennsylvanian rocks has been observed as far west as the western edge of the plateau in Van Buren, White, and Cumberland Counties. From Lake City, where the Jacksboro fault borders the Pine Mountain overthrust, to the Alabama state line, the structure along the eastern escarpment of the Cumberland Plateau is rather complex. Windows and **klippen** (the reverse of windows—outliers of older rock surrounded by a thrust fault and younger rocks) attest to the low angle of faulting and the upward ramping of the glide zones. Near Harriman, Tennessee, where faulting is very complex, one fault can be traced northwestward along the Emory River. This fault, like its apparent mirror image, the Jacksboro fault, northeast of the Wartburg Basin (see chapter 11), merges with a series of northeast-southwest trending faults in Morgan and Cumberland Counties. The similarity does not end there. Whereas the Pine Mountain overthrust sheet contains the Powell River anticline, the southern counterpart contains the Sequatchie anticline.

The Sequatchie anticline begins in the Crab Orchard Mountains south of the Emory River fault zone and extends for more than 200 miles to the southwest through Sequatchie Valley into Alabama. The anticline can be interpreted as a drag fold in the hanging wall of the upward ramping décollement. The northwest flank of the fold has been breached by the Sequatchie Valley thrust fault, where rocks of the Knox or Chickamauga Groups are thrust onto mostly Mississippian rocks. In Tennessee, rocks of the Knox Group are the oldest rocks exposed in the core of the fold.

Higher and higher incompetent stratigraphic units above the Rome Formation, such as shale units of the Conasauga and the Chattanooga Shale, serve as glide zones for faults during the westward tectonic transport. Finally, as is the case west of Sequatchie Valley, Pennsylvanian strata are faulted upon Pennsylvanian strata with glide zones lying mainly within the Richland or Sewanee coal seams of the Crab Orchard Mountains Group. Figure 13.6 shows cleavage in a coal seam and also how the seams thicken and thin as the upper unit glides across the lower unit causing pinching and swelling of the coal.

14

Breakup of Pangea

> He putteth forth his hand upon the rock; he overturneth
> the mountains by the roots. He cutteth out rivers among
> rocks; and his eye seeth every precious thing. He bindeth the
> floods from overflowing; and the thing that is hid bringeth
> he forth to light. But where shall wisdom be found? Where is
> the place of understanding?
>
> —Job 28:9–12

Pangea was the first supercontinent to be theorized. The name Pangea is taken from the ancient Greek, *pan* meaning "entire" and *Gaia* meaning "Earth." The term was first used during a 1927 symposium when Alfred Wegener's theory of continental drift was discussed. Based mainly, though not solely, on the jigsaw-puzzle-like shapes of continents, Wegener's theory postulated that at some point in Earth's history, before continents moved to their present locations, many were sutured together to form a large continent that Wegener called Pangea. Of course, we now know that there were other supercontinents that preceded Pangea (e.g., Rodinia) during the deep time of earth history (see chapters 6 and 8). The overwhelming proof of the existence of Pangea was a seed in the development of the great unifying geologic theory of plate tectonics.

Pangea existed as a supercontinent from the later part of the Paleozoic Era through the early part of the Mesozoic Era, a span of nearly 100 million years. Pangea was built gradually along its eastern or proto–North American margin through collisions with smaller terranes until finally colliding with the large proto-African continent. The uplifting that ensued from collisions along Pangea's margin caused deformation and erosion of all previously deposited rocks. Sediment that eroded from the uplifted mountainous areas was subsequently deposited to form the geologic record over parts of what is now Tennessee. For example, the Pennsylvanian strata in eastern Tennessee are products of the recycled sediments—truly, the rock cycle at work.

Pangea broke apart much like its predecessor, the supercontinent Rodinia, had done (about 750 Ma). Pangea began to rift apart about 180 Ma, separating the continents of North America and Africa (see figure 6.4). The major rift basin today is within the Atlantic Ocean; however, at the onset of the breakup, many smaller rift basins formed along the trend of a developing passive continental margin parallel to the present Appalachians. The detachment from Africa was not complete, as fragments of the African continent lie buried beneath younger rocks of Georgia and Florida. Although no rift basins are known in Tennessee, their presence is preserved in the bedrock of what is now the Piedmont province east of the Blue Ridge. Plate separation continues today along the divergent boundary or "spreading center," the Mid-Atlantic Ridge, with North America migrating in a northwesterly direction (see figure 3.8B).

Except for the extreme west, where the last vestiges of an ocean environment are recorded in the sedimentary record, Tennessee has remained a terrestrial environment. During this time the higher elevations were being incised and eroded by rivers and streams that carried their sediment loads to the passive margin of the Atlantic coast or to an insipient trough developing along the Mississippi River Valley—the New Madrid rift zone. So, although little in the way of fossils has been discovered in the eastern two-thirds of Tennessee (seldom are fossils found in terrestrial environments), this part of the state may have been a playground for dinosaurs during the Mesozoic Era, the "age of dinosaurs." Chapter 15 discusses in a bit more detail the evolution of Tennessee's landscape.

15

The Last 250 Million Years

> If by some fiat I had to restrict all this writing to one sentence, this is the one I would choose: The summit of Mount Everest is marine limestone.
>
> —John McPhee, *Annals of the Former World* (1998)

Introduction

During the long **deep-time** (tens of millions of years) history of the Appalachian region of Tennessee, a landscape resulted that was responsible for molding a unique socio-political, economic, and cultural state of being. A foundation of rock formations became buckled and ruptured during periodic collisions of tectonic plates (geosphere) and was finally sculpted by wind, water, ice, and organisms, including humans (atmosphere, hydrosphere, cryosphere, and biosphere) to form a spectacular landscape. Both the beauty and the harshness of the mountains, plateaus, gorges, ridges, and valleys wrought through time have strongly affected the music, politics, economics, customs, food, and many other aspects of life that have evolved in the Appalachian region. Over the course of time, the three grand divisions of Tennessee began to develop.

Paleogeologic Setting

Paleogeology (ancient geology) refers to what geologists would have observed at any point in the geologic past. For example, if you would have been walking almost anywhere across Tennessee around 250 Ma, the geology about you would have included outcrops of late Absaroka Sequence rocks (mostly Pennsylvanian age) in a countryside that was anything but flat. There would have been a chain of mountains on the distant horizon extending from what is now East Tennessee (the Appalachians) down to the southwest and then westward across Mississippi and Arkansas (what is today the Mississippi River valley) into Texas as the Ouachita and Marathon Mountains—all products of the Alleghanian orogeny.

Two other highland features were the Nashville Dome near the middle of the state and another off to the northwest called the Ozark Dome. Those plus another high area possibly related to the Ozark Dome, the Pascola arch, extended southeast-ward from the Ozark Dome across Missouri and West Tennessee. These highland features influenced the deposition of sediment in western Tennessee.

An intriguing explanation for the origin of the uplifted area referred to as the Pascola arch has been offered by Van Arsdale and Cox (2007). It involves the North American plate moving across a mantle plume called the Bermuda hot spot. Their supposition is that the rate of motion of the North American plate was just right for the Mississippi Embayment to pass over the Bermuda hot spot during middle Cretaceous time, the appropriate time for the uplifted region to be present. This part of the North American plate was already a weakened crustal region as a result of the rifting associated with the breakup of the supercontinent

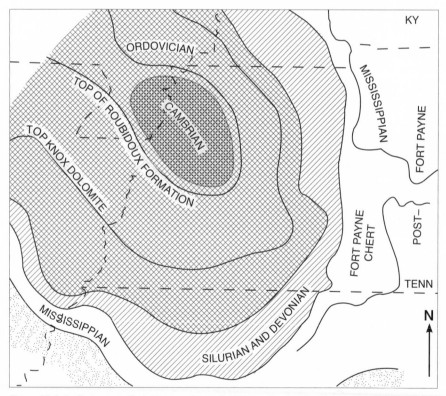

Figure 15.1. An Early or pre-Cretaceous paleogeologic map showing the distribution of rock units following uplift and subsequent erosion in West Tennessee and adjacent areas of the Mississippi Embayment. (Stearns, Marcher 1962.)

Rodinia (see chapter 9). Igneous intrusions in Arkansas and Mississippi possibly related to mantle plume activity have been dated between 120 Ma and 80 Ma (middle Cretaceous). A possible sequence of events according to this hypothesis might be as follows: (1) igneous intrusions were injected deep within the crust as the continent moved over the hot spot; (2) the thermal conditions associated with the hot spot caused the crust to be warped upward; (3) erosion removed thousands of feet of strata from the uplifted region, forming an outcrop pattern much like that shown in figure 15.1; (4) as the plate moved beyond the hot spot, material cooled, contracted, and subsided, aided by the presence of preexisting faults associated with the earlier mid-continent rifting of late Neoproterozoic. The subsidence ultimately allowed the incursion of the Cretaceous sea into the embayment. Figure 15.2 shows the arching of Paleozoic strata and their subsequent subsidence.

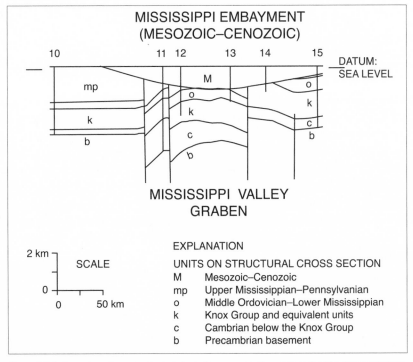

Figure 15.2. A west-northwest oriented geologic cross-section across the Mississippi Embayment south of Memphis illustrating the arching of Paleozoic strata (fig. 15.1) and subsequent subsidence of the rift, or graben. (Thomas 1991.)

Less obvious than the domes and folded and faulted rocks of the Appalachian and Ouachita mountains and influencing the development of Tennessee's landscape, there existed large-scale zones of fracturing wrought during the breakup of Pangea. These zones of fractured bedrock form weaknesses in bedrock, often creating preferred routes for rivers and streams (drainage patterns). Examples where fracturing may have controlled the course of rivers or streams include, of course, the channel of the Mississippi River that flows along the Reelfoot Rift of West Tennessee and the Pigeon, Hiwassee, Little Tennessee, and French Broad Rivers, which have incised nearly parallel channels through the Blue Ridge Mountains in East Tennessee. Other notable examples where fracture zones probably influenced the course of a river include the Emory River in Roane and Morgan Counties, affected by a strike-slip fault separating the southern Cumberland Plateau and the Cumberland Mountains (the Wartburg Basin), Sequatchie River, and possibly the many kinks or meanders in Tennessee River's contorted course.

The Rock Record

The last pages of Tennessee geologic history are found in the sedimentary record of the Zuni and Tejas stratigraphic sequences in the western part of the state. The Zuni Sequence records history beginning near the end of the Triassic Period and the beginning of the Jurassic Period, at about the time when the supercontinent Pangea was beginning to break apart. The Tejas Sequence follows the Zuni from the Cretaceous-Tertiary (K-T) boundary to the present. Current stratigraphic nomenclature has changed the name Tertiary Period to the Paleogene Period, thus it is more up to date to refer to the K-P boundary.

To a large extent, the present-day Mississippi Delta region is a modern analog for the geology of the Coastal Plain and Mississippi River valley of West Tennessee. Sedimentary deposits representing river floodplains, delta plains, barrier beaches and bars, lagoons, and deep ocean, all parts of the modern lower Gulf Coastal Plain, appear to be represented in the Zuni and Tejas sequences of Tennessee. The discussion herein of the complex sedimentary facies involved in the geologic history of western Tennessee relies very heavily on geologic investigations by Marcher and Stearns (1962); Stearns and Marcher (1962); Russell and Parks (1975); Stearns, Blythe, and Hoyal (1989); Gibson (1988 and 1990); and Van Arsdale (2009). However, despite their contributions, there remains a paucity of available geologic data for West Tennessee. Because West Tennessee has not received the same amount of attention as other regions of the state, West Tennessee is a true frontier for the quest of geologic knowledge.

Zuni Sequence

The sediments representing the Zuni Sequence in Tennessee lie almost exclusively in the West Tennessee uplands, a highly dissected hilly region that forms the divide between the Tennessee and Mississippi Rivers. The scarcity of a sedimentary record during Zuni time suggests that erosion was the dominant geologic process during that time. There is no known sedimentary record in Tennessee representing Late Pennsylvanian, Permian, Triassic, or Jurassic time. It was not until late Cretaceous time during the formative stages of the Mississippi Embayment in West Tennessee that sediment was deposited to create a geologic record. This record begins with patches of cherty gravel called the Tuscaloosa Formation that lies upon a deeply eroded surface of Paleozoic age rocks (primarily Devonian and Mississippian). At some locations the gravel appears to be fillings of sinkholes formed in the Devonian or Mississippian carbonate rocks, creating a **paleokarst** (ancient karst) on the pre-Cretaceous land surface. The gravel is a product of the long period of uplift and erosion extending over most of the Mesozoic Era from late Paleozoic time through late Cretaceous time. The Zuni Sequence closed with marine and nonmarine sediments deposited upon the Tuscaloosa Formation as the Cretaceous sea ebbed and flowed within the Mississippi embayment.

Figure 15.3 is a **paleogeologic map** (a map drawn to show the distribution of rocks at the land surface at a particular point in time in the geologic past) of pre-Cretaceous time in West Tennessee. Note that the ancient outcrop pattern with the oldest rocks in the center indicates an uplifted area. This uplifted area is the Pascola arch—interpreted to be a significant source of the Tuscaloosa gravel. It has been estimated that, even though nearly 8000 feet of strata had been stripped from the crest of the arch prior to the beginning of Tuscaloosa time, the arch remained a formidable topographic feature standing at least 1000 feet above sea level and sloping eastward to an ancient coastal plain that existed in the vicinity of the present Western Highland Rim of Tennessee.

The Tuscaloosa Formation has two facies—a western facies interpreted as alluvium deposited across the karstic plain of the Western Highland Rim by a network of eastward flowing streams heading on the Pascola arch, and an eastern facies that was deposited east of and downslope of the western facies along the beach of an encroaching Cretaceous sea. The stream deposits of the western facies are poorly sorted and consist mostly of chert pebbles and cobbles (some up to six inches in diameter), suggesting rather steep stream gradients; whereas, in the eastern facies, the deposits are better sorted, which is what would be expected if the sediment was being winnowed by waves and long-shore currents, and contain more quartz pebbles and sand as well as heavy minerals (typically

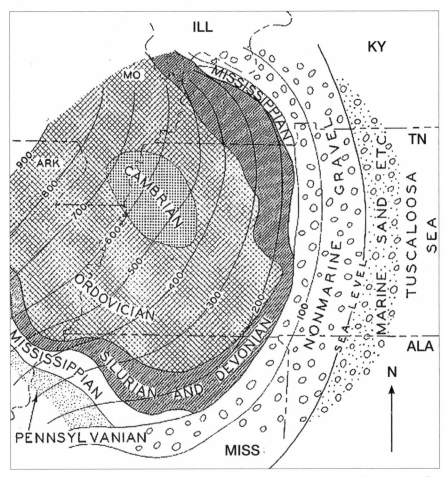

Figure 15.3. An Early Cretaceous paleogeologic map of western Tennessee and adjacent areas of the Mississippi Embayment showing the general location of facies of the Tuscaloosa Formation. Contours show the relief and trend of the Pascola arch. (Marcher, Stearns 1962.)

magnetite, rutile, and zircon), which might possibly have been washed in from source areas to the north or south.

Disconformably overlying the Tuscaloosa gravels are transgressive marine sands called the McShan and Eutaw Formations. Facies of **glauconitic** sands (glauconite is a green micaceous mineral formed by biochemical alteration of iron-bearing minerals like biotite typically found in sediments formed in shallow marine environments) and very thick-bedded sands that may be **conglomeratic** (often called Tombigbee Sand) were probably deposited in a nearshore environment as the Cretaceous sea transgressed. The overlying formations, the Coffee,

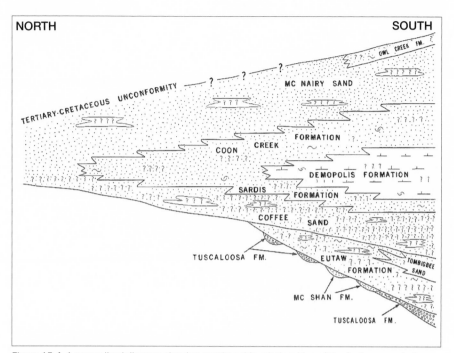

Figure 15.4. A generalized diagram showing stratigraphic relationships of the Cretaceous strata of western Tennessee. Note the relationship of the Demopolis Formation with other Cretaceous formations. (Russell, Parks 1975.)

Sardis, Demopolis, Coon Creek, and McNairy, are closely related sedimentary facies of a thick sequence representing a transgressive-regressive-transgressive cycle. Figure 15.4 illustrates the stratigraphic relationships among the Cretaceous strata of Tennessee.

The Coffee Sand, which ranges in thickness up to 200 feet in McNairy County and is composed mostly of stratified micaceous sand with thin lamina of clay containing woody fragments or lignite, has been interpreted to represent barrier beaches and lagoons of an advancing "Demopolis Sea." Fossils in the Coffee Sand are rare. The Sardis Formation is a glauconitic sand that overlies and grades into the Coffee Sand and is interpreted as a continuation of the transgressive sequence. The Sardis is characterized by its abundance of green glauconite that, on weathering, develops into a distinctive yellowish-orange color from oxidation of the iron in the glauconite. As seen in figure 15.4, the Demopolis Formation forms the core of a transgressive-regressive wedge of Cretaceous sediments. Figure 15.5 shows a probable paleoenvironmental interpretation of the Demopolis Sea prior to its regressive phase.

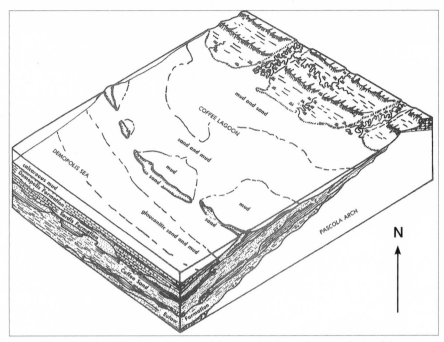

Figure 15.5. A hypothetical paleoenvironmental diagram showing sedimentary facies of the Demopolis Sea. (Russell, Parks 1975.)

The Demopolis, consisting of marls, chalky marls, and calcareous clays, is considered to have been deposited offshore on a relatively deep shelf. The nearer to shore, shallow marine sands such as the Sardis Formation grade laterally into the Demopolis (see figure 15.4). Lithologically, the Demopolis is a massively bedded fossiliferous, shaly, gray marl. Though no true chalk occurs in Tennessee, the Demopolis is probably as close as it gets. In places the fresh rock is tough but weathers to a **conchoidally fracturing** (curved fractures like broken thick glass or an old Coke bottle), light-gray, chalky marl. The Demopolis is estimated to range up to 90 feet in thickness. *Exogyra ponderosa* and *Exogyra costata* are among the characteristic species of bivalves occurring in the Demopolis.

The Coon Creek Formation is interpreted to represent the initial deposit during the regressive phase of the "Demopolis Sea" (see figure 15.5). It was deposited in a more or less mixing zone between the deeper marine sediment of the Demopolis and the nearshore and fluvial (river) deposits of the McNairy Sand. The close of the Demopolis Sea is marked by a thick sequence of sand, the McNairy Sand. At the base is a fine-grained sand about 50 feet thick containing abundant heavy minerals indicating a nearshore environment as the sea began to retreat.

As the sea continued to regress, coarser sands were deposited upon the fine sand in a transitional environment with rivers, shallow bays, and deltas between the land and the sea. The upper sand of the McNairy possibly represents a nearshore marine environment deposited as a last impulse of the sea before final withdrawal at the end of Cretaceous time and the end of the Zuni Sequence.

Tejas Sequence

The break between the Zuni and Tejas sequences is a profound one that is recognized the world over. It is referred to as the K-T or K-P boundary, Cretaceous-Tertiary or Cretaceous-Paleogene, respectively, that marks a dramatic change in the biota of the planet—most commonly thought of as the demise of the dinosaurs. Current geologic time nomenclature now refers to the Tertiary as the Paleogene Period. The mass extinctions associated with this interval of geologic time are thought to be related to the onslaught of meteorite impacts on the earth. The most significant impact site is Chicxulub on the Yucatan Peninsula in Mexico. Because the fallout from this gigantic impact has been recognized around the world, there is little doubt that sunlight was blotted out by an atmosphere that must have been laden with debris. Other than the impact on terrestrial life, the only other likely impact from Chicxulub might have been a gigantic tsunamis rolling up the Mississippi Embayment.

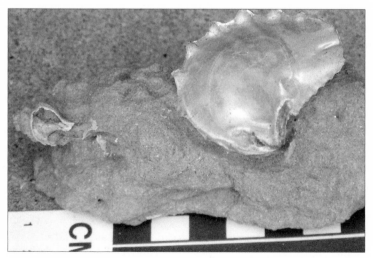

Figure 15.6. *Pterotrigonia (Scabrotrigonia) thoracica*, "Ptero," the Tennessee state fossil. The fossil is preserved as original shell material in the world famous Coon Creek locality in McNairy County, Tennessee.

The sedimentary record of the Tejas, as a famous baseball player, Yogi Berra, would say, is "*déjà vu* all over again." During the Paleocene through Miocene Epochs, mostly unconsolidated sedimentary deposits similar to the Zuni Sequence, formed sedimentary facies of shallow to deep marine conditions to nearshore **fluvial** (river) environments, deposited as the ocean transgressed and regressed through the Mississippi Embayment.

The return of a shallow sea to West Tennessee is marked by the deposition of the Paleocene Clayton Formation and the Porters Creek Clay of the Midway Group. The Clayton Formation, the basal unit of the Tejas transgressive sequence, in many respects resembles the McNairy Sand deposited during the regressive phase of the Zuni Sequence. Both were deposited in a nearshore environment, and it is likely that much of the sandy portions of the Clayton Formation consists of reworked sediment from the McNairy. One notable difference from the McNairy is that limestone occurs within the lower Clayton. The Clayton averages 80 feet in thickness. Overlying the Clayton Formation, and representing a deeper marine environment, is the Porters Creek Clay. The Porters Creek is typically a thick body of brownish-black, olive-black, grayish-black, and dusky yellowish-brown clay averaging 100 to 150 feet in thickness. Lenses, dikes, and interbeds of sand occur within the Porters Creek and may possibly be related to past earthquake activity (sand blows resulting from liquefaction caused by seismic shock waves).

A disconformity between the Porters Creek Clay and the overlying Wilcox Formation suggests that there was a withdrawal of the sea and a time of erosion before the Eocene Wilcox Formation was deposited. The Wilcox may be up to 200 feet in thickness and consists of a heterogeneous mix of quartzose sand, silt, and clay, none of which has any lateral continuity. The Wilcox contains minor beds of lignite, kaolin, and siderite (iron carbonate), but most distinctive of the formation is the presence of ball clay conglomerates and the oft-referred-to "sawdust sand" (appearance of weathered outcrops).

Overlying the Wilcox Formation, or in some cases the Porters Creek Clay where the Porters Creek overlaps the Wilcox, is the Claiborne Formation. The Claiborne is predominantly composed of the nearly 300-foot-thick Memphis Sand that forms the major aquifer for water supplies in western Tennessee, eastern Arkansas, and northwestern Mississippi. The depositional environment for the Claiborne Formation was likely that of streams flowing across a low coastal area with interspersed lakes, swamps, and lagoons. The presence of freshwater lakes and swamps has been interpreted from the deposits of ball clay containing fossil flora (see figure 15.7). The commercial deposits of clay are referred to as ball clay because of the shape of the clay when processed for shipping. The clay deposits, ranging in thickness from less than a foot up to 50 feet, are not continuous, making prospecting for economical deposits difficult. A test well near Fort Pillow, Tennessee, recorded a total thickness of 1100 feet for the Claiborne.

Overlying the Claiborne Formation is a very poorly exposed unit called the Jackson or Cockfield Formation of Eocene-Oligocene age. The formation is mostly silty sand or silty clay and probably represents the last time an ocean extended into the upper parts of the Mississippi Embayment.

With the exception of being several hundred feet higher in elevation than today, it is estimated that, by early Pliocene time (~5.3 Ma), the Mississippi River Valley (or West Tennessee Plain) looked much like today, with a major river flowing down the middle within a wide floodplain. Remnants of this stage in Tejas history consist of discontinuous patches of sand, silt, and clay perched on hilltops.

From about 1.8 Ma to about 10,000 years ago—the Pleistocene Epoch—North America was besieged by at least four major glacial ages, with ice advancing at times as far south as the Missouri and Ohio Rivers. As each glacial episode waned, huge volumes of water transported glacial outwash down the Mississippi River valley. There were probably times when the sediment load was so great that there was insufficient volume of water to transport the material freely, and a braided river resulted, with islands separated by converging and diverging channels. Westerly winds deflated the fine sediment from the braided river islands, transporting it to the east and forming thick deposits of löess.

Löess is unique in the sense that even though it is essentially an unconsolidated deposit, it has the propensity to stand in vertical embankments. The frosted, somewhat elongate grains interlock, allowing it to stand at an atypical

Figure 15.7. Fossil leaf in clay of the Claiborne Formation collected from a ball clay quarry in Weakley County, Tennessee.

Figure 15.8. Löess bluff near Reelfoot Lake in Obion County, Tennessee. Because of the texture of the loess, exposures are most stable when vertical.

angle of repose. Early geologists described the silt as being **fluvial** (river) in origin, but the discovery of terrestrial forms of snails clearly suggests **eolian** deposition (silt on a land surface deposited by wind). The löess, perhaps up to 200 feet thick, forms the steep Chickasaw Bluffs overlooking the Mississippi alluvial plain. Figure 15.8 is an outcrop of löess and figure 15.9 shows a fossil snail in the löess.

The Landscape

Perhaps as interesting as the formation of the rock record during this time is the work by the Earth's systems in creating the spectacular scenery of Tennessee's landscape. While layers of sediment were recording the pages of history in the westerly portions of Tennessee (Zuni and Tejas Sequences), pages of earlier history were being removed from the record as the land was being sculpted by erosion over most of the remaining parts of Tennessee. Nearly all of Tennessee's scenery is explainable by the interactions of the geosphere, hydrosphere, atmosphere, biosphere (including us), and cryosphere during the most recent geologic history of Tennessee, including the enigmatic course of the Tennessee River, flowing in every direction of the compass, as it flows from its headwaters in the ridges of

Figure 15.9. Fossil snail in löess bluff shown in fig. 15.8.

East Tennessee to its confluence with the Ohio River near Paducah, Kentucky. Because it is not in the purview of this book to discuss details of Tennessee's landscape, the reader is directed to other resources available that describe and explain in detail specific topographic elements, such as waterfalls, natural bridges, caves, and other features.

A great deal of what we see, especially in East Tennessee, is what is known as "inverse topography." This is where originally structurally formed uplands, such as anticlines and domes, are converted into lowlands through erosion and structurally formed lowlands, such as synclines, become uplands. In the early stages, up arched anticlinal and domed areas such as the Sequatchie anticline and the Nashville Dome, respectively, formed uplands, and down warped synclinal areas such as Lookout Mountain near Chattanooga or Bays Mountain in northeast Tennessee formed lowlands. But as erosion-resistant rocks of the upland areas were breached, more easily eroded rock layers became exposed, and the anticlines and domes were gradually reduced to form lowlands such as the Sequatchie Valley and the Central Basin of Nashville. In a similar fashion, the same erosion-resistant layers that once capped the anticlines and domes are eventually encountered as the original lowlands are eroded, forming a cap for a ridge or mountain. The resistant layers result in a condition of dynamic equilibrium where ridges and mountaintops form at correlative elevations determined by erosional resistance of certain layers of rock. Figure 15.10 illustrates the evolutionary steps of how erosion creates topographic inversions as structural high areas are transformed into topographical low areas through dynamic equilibrium. Thus today the Sequatchie anticline forms a long valley; Lookout Mountain, a synclinal area, is a mountain; and the Nashville Dome is a basin.

Waterways

The waterways of Tennessee, especially the Tennessee River, have always been important to the state's economy. A study of the configuration of the waterways in Tennessee is interesting in of itself (see figure 15.11). Fold axes and fault traces in the Valley and Ridge province of eastern Tennessee provide a northeast-southwest grain of alternating rock types with varying resistances to erosion. Large rivers and streams, mostly tributaries to the Tennessee River, tend to flow either northeast or southwest through the more easily eroded rocks of the northeast-southwest oriented grain. The tributaries to these streams and rivers head on the adjacent ridges of more resistant rock and join at nearly right angles. When this arrangement is viewed from above, the longer rivers and streams in the valleys with tributaries joining at right angles create a pattern resembling a trellis; hence, the drainage pattern is referred to as trellis.

On the other hand, where strata are essentially horizontal and equally resistant to erosion in all directions, the flow of a river or stream is mainly controlled by the slope of the land. The result is a drainage pattern called dendritic that resembles the branches of a tree. The minimal rock deformation and the rock layers being mostly horizontal from the Cumberland Plateau westward mean that dendritic drainage dominates. The outcrop pattern on a Tennessee geologic map

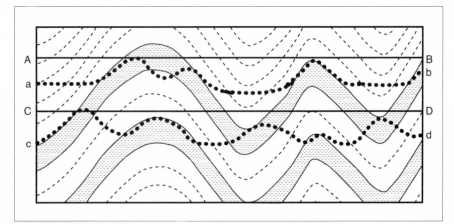

Figure 15.10. A sketch illustrating inverse topography. Lines A–B and C–D represent time lines and a–b and c–d represent the topography at those points in time. The stippled layer is an erosion resistant layer. Note the various landforms developed as erosion lowers the surface with time.

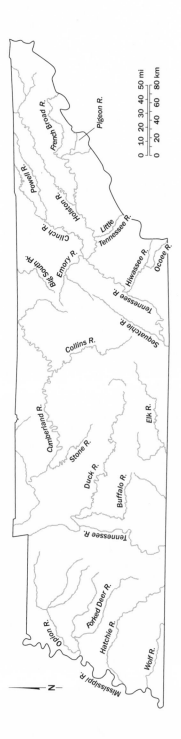

Figure 15.11. A sketch map of major waterways in Tennessee.

where streams are eroding headward into the Eastern Highland Rim and the western escarpment of the Cumberland Plateau displays an excellent example of a dendritic pattern.

The Tennessee River itself is an entirely different matter. With its headwaters on the east side of Knoxville at the confluence of the Holston and French Broad Rivers, the Tennessee flows southwesterly through the valley of East Tennessee, forming water gaps through many ridges and then turning westward at Chattanooga, forming a loop through Alabama and Mississippi only to flow northward back into and through West Tennessee, to join up ultimately with the Ohio River. Because of this lengthy course, it has always been a strategic route for commerce. During the Civil War, its control was critical, and thus many battles centered on the ability to control its traffic. During the Revolutionary and Civil Wars many ridges became redoubts to establish control over water as well as land transportation routes.

Because of their deep valleys, many Tennessee River tributaries heading in the Blue Ridge have been impounded in order to generate hydroelectric power. Dams were constructed on the Little Tennessee River and some of its tributaries to generate power for an aluminum industry developing in Alcoa, Tennessee. Also, since the inception of the TVA (a federal program instated in the 1930s for the primary purposes of flood control, generation of electricity, navigation, and recreation), the entire length of the Tennessee River is a sequence of lakes impounded by dams.

The other major river in Tennessee is the Cumberland River, which heads in Kentucky, enters Tennessee in Clay County, skirts westward toward Nashville along the northern portion of the state, and reenters Kentucky as Lake Barkley in Stewart County before ultimately flowing into the Ohio River. Much as with the Tennessee River, a number of dams have been constructed on the Cumberland by the U.S. Corps of Engineers for flood control, recreation, and power generation.

The Tennessee River and a few other major rivers do not exactly fit into either the trellis or dendritic category. Although the Tennessee River course in East Tennessee follows the grain of the Appalachian structure in general, it is apparent that the river indiscriminately eroded its channel across rocks of different resistances on its route southwestward to the vicinity of Chattanooga. At Chattanooga the river seems to defy what should be a natural route through the Coosa Valley to the Gulf of Mexico and instead takes a meandering path through Walden Ridge of the Cumberland Plateau. Also, rivers such as the Pigeon, Hiwassee, Little Tennessee, and French Broad appear to have taken the hard route through resistant rocks and cut gorges through the igneous and metamorphic rocks of the Blue Ridge Mountains. The "64 thousand-dollar-question" is why.

The Tennessee River course is one of the most enigmatic features of the present landscape of Tennessee. The river flows approximately 652 miles in a very convoluted course that includes every direction of the compass on its way from East Tennessee to the confluence with the Ohio River in Kentucky. What is so interesting is that after the river flows south through the valley of East Tennessee to a point near Chattanooga, the river abruptly turns its course to the northwest where it flows through a gorge in Walden Ridge forming the "Grand Canyon" of the Tennessee River (see figure 15.12). Why does the river turn to the northwest here when it appears that it could have continued to flow southward to the Gulf of Mexico via the Coosa Valley? Once through Walden Ridge, the river flows along the trend of the Sequatchie Valley and would seem to have another clear shot to the Gulf through the Black Warrior River Valley, but instead near Guntersville, Alabama, the Tennessee River flows again to the northwest. Then, as the river nears the Alabama-Mississippi state boundary, it turns northward, entering Tennessee again, and finally it flows into the Ohio River near Paducah, Kentucky.

A paper by Mills and Kaye (2001) presents an excellent review of hypotheses dating back to 1875 that have been offered to explain the peculiar course of the

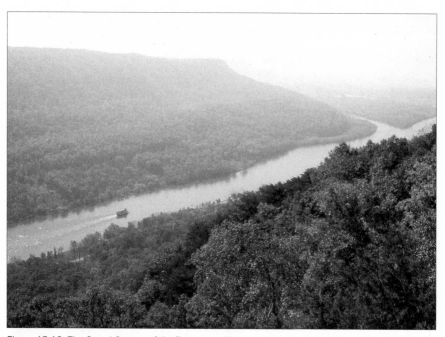

Figure 15.12. The Grand Canyon of the Tennessee River as it flows westward below Signal Point, Signal Mountain, Tennessee.

Tennessee River. They conclude that though many geologists have made interesting observations of geologic data, possibly related to deciphering the puzzle, these data are complex and confusing, rendering any solution in the future very difficult. In the final analysis, it is likely to be concluded that the course is attributed to a number of factors. For example, **stream piracy,** a process whereby one stream or river having a more favorable **gradient** (a slope offering an easier route to base level) captures another stream or river with a less favorable gradient, might have been involved in some instances. One example where stream piracy has been offered as an explanation is at Chattanooga. This proposal suggests that a river related to the drainage in Sequatchie Valley could have been eroding headward toward the east across what is now the Cumberland Plateau, intersecting and capturing the Tennessee River and preventing it from flowing down the Coosa Valley in Alabama. Uplift of the land is certainly another likely contributor to the Tennessee River course, especially as the river skirts to the northwest and then north around the Nashville Dome.

Waterfalls

Waterfalls are among most scenic features in the landscape of Tennessee. Waterfalls result when rocks resistant to erosion are encountered in the course of a river or stream, causing the water to descend vertically. As the resistant rock is gradually worn down, the waterfall eventually morphs into a cascade or series of small waterfalls. Both falls and cascades are especially common in the Blue Ridge, Valley and Ridge, Cumberland Plateau, and Highland Rim sections of Tennessee. Dunigan (2011) cites more than 643 waterfalls in Tennessee, and Plum (2009) provides excellent descriptions with directions to the more than 650 waterfalls noted in Tennessee.

In the Blue Ridge, most of the waterfalls cascade over the resistant metamorphosed rocks of the Great Smoky Group. The author's favorites in the Blue Ridge include Laurel, Abrams, Rainbow, and Grotto Falls, as well as Ramsey Cascades in the Smokies and Bald River and Benton Falls in the Cherokee National Forest. On the Cumberland Plateau, massive beds of sandstone capping more easily eroded shale are the cause for most waterfalls. The most famous of the plateau waterfalls, at 256 feet, is Fall Creek Falls, located in Van Buren County within the State Park bearing that name. Cane Creek Falls, also in Fall Creek Falls State Park, Ozone Falls in Cumberland County, and Foster Falls in Marion County are a few others on the author's list of favorites. However, Virgin Falls in White County also finds a place on the list because of its spectacular features. Virgin Falls developed on the western edge of the Cumberland Plateau where erosion has incised below the sandstone cap of the plateau into limestone that is notable for underground drainage associated with cave development. It is not uncommon for streams flowing westward off the plateau to disappear by losing their surface water into the underlying cavernous limestone. The stream at Virgin Falls emanates from a cave, flows overland for about 20 yards, and then drops about 100 feet into another cave. The other section in Tennessee where there is a significant change in elevation is between the Highland Rim and the Central Basin. Perhaps the best example of a waterfall on the rim is Burgess Falls, located in a State Natural Area on Falling Water River.

Interesting Landforms

Rock deformation, the relative resistance to erosion of rocks, and subsequent erosion has sculpted a number of interesting bedrock features such as natural bridges, arches, and other monoliths, most of which have been dubbed with special names. Many of these features provide some of the most scenic spots in Tennessee.

Structural elements such as rock joints (see chapter 3) control the erosion that is the major process creating the interesting landforms of Tennessee. Jointing is a major factor in the formation of natural bridges, arches, and other forms that might be regarded as stand-alone monolithic features. The majority of these features are found on the Cumberland Plateau, where the bedrock is flat lying and joints perpendicular to bedding tend to divide the rock into massive blocks. The most common term for a monolithic feature is probably *chimney*, but the use of *devil* as in devil's shinbone, slide, or backbone is common. The vertical columns of rock, or chimneys, typically bounded on all sides by vertical joints, became separated from a rock face such as a cliff or bluff by water action, freezing and thawing of ice, the growth of tree roots, or a combination thereof taking place within joints. A feature that piques the curiosity of many who pass it by on I-75 in Campbell County is the Devil's Racetrack, a set of prominent vertical columns of sandstone exposed on the southwestern end of Cumberland Mountain (see figure 1.10).

Natural bridges and arches are among the most attractive scenic erosional features in Tennessee's landscape. As spans or arches of rock that resemble human-made bridges, they have captivated the people's imaginations for centuries. They can be found on ridge crests and in valleys and formed in sandstone or limestone, but they are always related to rocks having resistance to erosion. The majority of Tennessee's bridges and arches are associated with essentially flat-lying sandstone formations located on the Cumberland Plateau. As in the case of monolithic rock features, jointing plays an important role in the development of an arch or bridge. In the case of bridges or arches developing in a valley, joints in

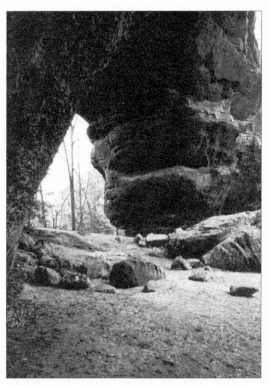

Figure 15.13. One of the Twin Arches in the Big South Fork National River and Recreation Area. Photo by Chuck Summers.

the bedrock of a stream channel create pathways for water flowing across the streambed to undercut its channel; then, as the stream continues to erode down along joints below the original streambed, part of the original streambed is left as a bridge or arch (see figure 15.13). Another scenario for the formation of natural bridges and arches, particularly those occurring along ridge crests, involves headward erosion by a stream (lengthening the stream valley by erosion) on both sides of a ridge. As the streams attack on both sides of a ridge, jointing causes rock falls, producing sand caves that continue to enlarge and ultimately forming a connection between the two sides with a span of resistant rock above.

Window Cliffs, on the Highland Rim in Putnam County and located about a mile and a half northwest of Burgess Falls Natural Area, are features not ordinarily included in the discussion of arches and bridges, but the windows are worthy of mention, given their scenic beauty and interesting mode of origin. Cane Creek, which heads in Cookeville, forms a tight meander bend, leaving a narrow ridge of horizontal siliceous Fort Payne Chert with a well-defined set of joints on the

Figure 15.14. Window Cliffs, Putnam County, Tennessee.

inside of the meander bend. Weathering of this ridge has loosened and removed blocks defined by horizontal bedding planes and vertical joints, leaving resistant lentils of Fort Payne to form rectangular shaped windows (see figure 15.14).

Arches and bridges in limestone result from a type of erosion different from sandstone. Limestone erodes mainly through dissolution, as evidenced by features such as caves, caverns, and sinkholes, all formed when limestone is dissolved. Jointing also plays a role in the erosion of limestone. **Karst** is a term used for the landscape features found in areas of limestone bedrock. Where limestone is the bedrock of an area, surface streams are often missing and underground drainage dominates. Many valleys are "dry valleys" except when large rainfall events supply water faster than it can be taken into the underground drainage system and flow as surface water. **Sinkholes,** depressions in the land surface formed by collapse into underground voids created by dissolution of limestone, springs, and dry valleys, are all characteristics of karst. If a stream flows into a sinkhole, it may emerge as a flowing spring below the sinkhole. Over time, erosion reduces the distance between the sinkhole and the spring, forming a span or natural bridge. Such a feature has formed above Virgin Falls described in the waterfall section above.

Karst Features

Limestone bedrock exists over a very large portion of Tennessee; therefore features of karst are common and widespread. The caves, caverns, and other erosional features associated with dissolution of limestone began forming soon after the breakup of Pangea and continue today. Some features already have been described in earlier chapters, such as the structural windows or coves such as Cades Cove in the Blue Ridge, Sequatchie Valley, and other valleys in the Valley and Ridge province, but there are several karst features that deserve special mention. Among them are Grassy Cove northeast of Sequatchie Valley, White Oak Sink in the Great Smoky Mountains National Park, and the Gray Fossil site in Washington County.

Grassy Cove, a very large valley completely enclosed by mountains and located about five miles east of Crossville, has been designated a National Landmark. The cove is a very large sinkhole that over time formed through the coalescence of many individual sinkholes developed in the Mississippian limestone. The drainage is internal, with no surface streams flowing from the cove. The cove floor has an average elevation of 1500 feet—about 600 feet higher than Sequatchie Valley to the southwest. The mountains bounding the cove are capped by Pennsylvanian sandstone and form some of the highest elevations on the Cumberland Plateau. The cove is surrounded by Brady Mountain on the west with an elevation of 2930 feet, Bear Den Mountain with an elevation of 2828 feet on the east, and by Black Mountain on the north. Brady Mountain and Bear Den Mountain converge to make the southern wall. Hinch Mountain, with an elevation more than 3000 feet, lies to the south of the Brady Mountain–Bear Den Mountain convergence and forms the northern boundary of Sequatchie Valley. Grassy Cove and Sequatchie Valley are structurally and genetically related. The anticlinal structure of Sequatchie Valley extends northeastward through Grassy Cove to Crab Orchard.

Grassy Cove, another example of topographic inversion, represents an early stage in the erosional process that created Sequatchie Valley. One scenario would be that the breaching of the sandstone cap of the Sequatchie Valley anticlinal structure migrated northeastward by headward erosion, assimilating preexisting coves similar to Grassy Cove as their sandstone caps were breached. Figure 15.15 is a diagram by Milici (1968A) illustrating the landform evolution of Sequatchie Valley and Grassy Cove. Parts D and E of the diagram show the development of an incipient Grassy Cove, which will eventually become part of Sequatchie Valley. Actually, the Sequatchie River heads in Grassy Cove. Grassy Cove Creek, the main surface stream in the cove, flows into Mill Cave (see figure 15.16), where it then flows through a labyrinth of underground cave passages ending at Devil's

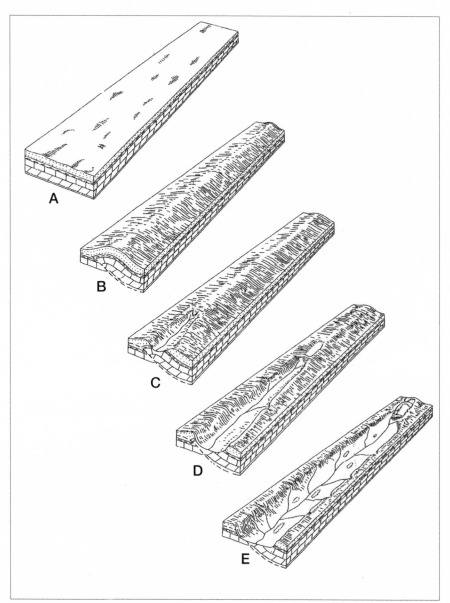

Figure 15.15. The evolution of Sequatchie Valley and Grassy Cove. A. At the close of Pennsylvanian time (~300 Ma), the region occupied by Sequatchie Valley was capped by sandstone. B. The region was folded (the Sequatchie anticline) and faulted during the Alleghanian orogeny. C. Erosion breaches the sandstone cap of the anticline exposing the underlying Mississippian limestone to underground erosion (karst development). D. The Sequatchie River lengthens its course by headward erosion through the easily eroded limestone. E. Grassy Cove represents the next part of the anticline that will be added to Sequatchie Valley.

Figure 15.16. Mill Cave in Grassy Cove receives most of the cove's drainage, which in turn flows through a cave system and becomes the headwaters of the Sequatchie River.

Step Hollow Cave in Brady Mountain, where it emerges as the head of Sequatchie Spring—the beginnings of the Sequatchie River.

White Oak Sink lies on the edge of Tuckaleechee Cove near the western boundary of Great Smoky Mountains National Park. Tuckaleechee Cove is a structural window like Cades Cove and other windows in the Blue Ridge formed when erosion breaches resistant rocks that are thrust upon easily eroded limestone. White Oak Sink on the edge of Tuckaleechee Cove represents ongoing erosion that is continually enlarging the cove. At the sink a waterfall cascades over resistant Neoproterozoic metamorphic rocks above the Great Smoky fault into a sinkhole in Paleozoic limestone (see figure 15.17). The water then flows beneath a dry valley to become part of some deep caves at the end of a "blind valley" and then, based on dye tracing, flows through the Tuckaleechee Caverns before emerging at a spring located at a trout-rearing farm almost two miles from the sink.

The Gray Fossil Site located near the town of Gray in Washington County is another example of topographic inversion. In this case, what was once low is now high. The site is a knoll sitting about 60 feet above the surrounding valley but is believed to represent a freshwater pond within an ancient sinkhole that existed prior to the erosion that formed the present valley (see figure 15.16). Fossils dated

Figure 15.17. White Oak Sink, Great Smoky Mountains National Park. The sink lies on the margin of Tuckaleechee Cove. The water is flowing over Precambrian rock on the hanging wall of the Great Smoky Mountain fault, and the water is entering a cave in the Ordovician Jonesboro Formation in the footwall of the fault. The student on the right is marking the location of the Great Smoky fault.

as late Miocene to early Pliocene (11.6–2.6 Ma) occur within layers of clay and silt that are black, gray, white, and buff (Shunk 2006). Much of the black material contains woody fragments of lignite. Based on a wide variety of vertebrate and invertebrate fossils discovered at the site, it has all the earmarks of a watering hole. Gravel composed of limestone fragments and chert, large blocks of limestone possibly representing sinkhole collapse, and chunks of calcite of possible cave origin are all features indicating a karst origin for the fossil site. Today the site is an example of inverted topography. Figure 15.18 shows a hill where once a depression existed as a watering hole for a variety of mammals and reptiles. Excavation of this hill continues at the site, and an excellent museum displays the flora and fauna collected. Figure 15.19 is an artist's rendition of what the Gray site might have looked like 11.6–2.6 Ma.

Beneath the bed of Dry Fork Creek near Spencer, in Van Buren County, on the Eastern Highland Rim is one of the largest caves in the eastern United States—Rumbling Falls Cave. Access to the cave is by rope down an 80-foot pit. The Rumble Room in the cave is estimated to be four acres. Figure 15.20 is a photograph of the pit and the Rumble Room taken by photographer-spelunker, Chris Anderson.

Figure 15.18. The Gray fossil site, Washington County, Tennessee. The hill in the center is an inversion of the original topography millions of years ago when the site was a pond in a sinkhole.

Other Interesting Features

Reelfoot Lake, adjacent to the Mississippi River in the northwest corner of Tennessee in Lake and Obion Counties, is a sunken area resulting from a series of earthquakes in 1811–1812. The earthquake was so powerful, it is alleged, that church bells as far away as Boston were caused to ring. Prior to the quake, the site was a typical floodplain with oxbow lakes and meander scars with cypress trees and other swampy vegetation. The quake caused the Mississippi River to reverse its flow, the land to warp and subside along faults, **sand blows** (eruptions of sand from the subsurface) to deposit mounds of sand, and slumping of the löess bluffs along the left bank of the river.

Most of Reelfoot Lake is a maze of swampy vegetation and tops of submerged cypress trees. Many trees are totally submerged below water level, but along the courses of streams that existed prior to the quake, there are meandering rows of cypress trees that were growing on old natural levees (see figure 1.12).

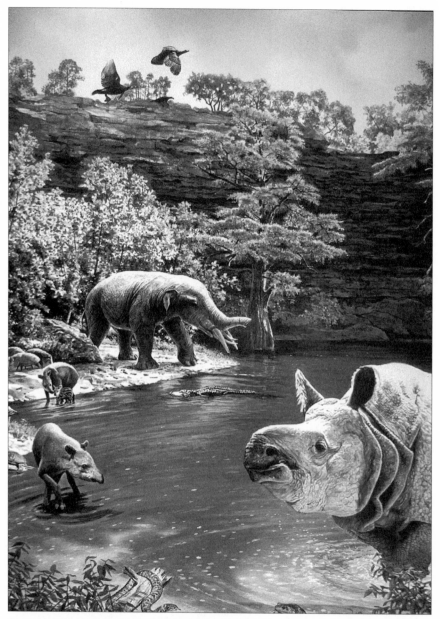

Figure 15.19. A mural in the McClung Museum, the University of Tennessee, depicting how the Gray Site might have appeared 4.5–7 Ma. The site is renowned for its diverse fauna. Portrayed here are a rhinoceros on the right, several tapir on the left, a swimming alligator just below an elephant in the background, a turtle at the water's edge, a snake in the foreground, and some circling birds. (Painting by Rob Wood, Wood Ronsaville Harlin, Inc. Courtesy of the McClung Museum, University of Tennessee, Knoxville.).

Figure 15.20. A view of the Rumble Room in the Rumble Falls Cave near Spencer, Van Buren County. For scale note the spelunker on the rope dangling through the pit and the spelunkers in the foreground. The Rumble Room is estimated to be 4 acres. (Courtesy of Chris Anderson).

Impact Structures

Meteors have bombarded the Earth since the time of its very beginning. One of the more famous impact structures, known for allegedly having a major role in the extinction of dinosaurs at the end of the Cretaceous, is Chicxulub in the Yucatan and is mentioned earlier in this chapter. Several impact structures, though not as great as Chicxulub, have been discovered in Tennessee. The appearance of Tennessee's impact craters is subtle. They are not obvious like Meteor Crater at Winslow, Arizona. The contrasting differences lie in the climate of the two regions. The area around Winslow is desertlike, where weathering and erosion are less destructive of surface features when compared to the warm and humid climate of Tennessee, where weathering and erosion can modify a land surface relatively fast. Also the Meteor Crater is much younger than Tennessee's.

The largest known of the Tennessee structures is the Well Creek Structure located south of Cumberland City on the Cumberland River athwart the county line between Stewart and Houston Counties. The structure was first noted when

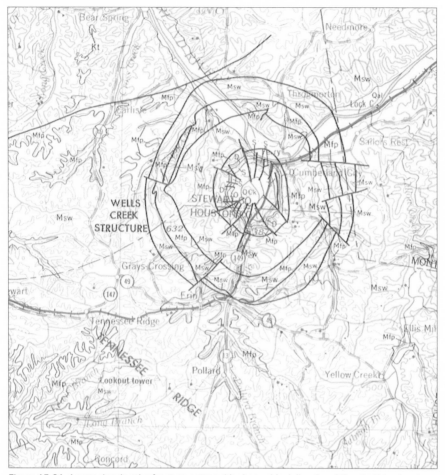

Figure 15.21. A map showing the fracture pattern (the heavy black lines) and the stratigraphy at the Wells Creek impact structure. Note that the Cambrian-Ordovician Knox Group is exposed in the center of the structure, and the rocks on the periphery are Mississippian in age. The much older Knox was brought to the surface by rebound after the impact.

chaotic blocks of rock were found during excavation for a railroad. Faulting resulting from impact defines a structure eight miles in diameter (see figure 15.21). No physical remains of a meteorite have been discovered, nor have there been detected any remnant magnetic signatures suggesting presence of metallic remains. Chaotic blocks, fine **brecciated** (fragmented) rock, faulting, and spectacular specimens of **shatter cones** (conically shaped masses of rock formed by high impact) provide the best evidence for impact from an extraterrestrial source.

Features analogous to shatter cones are observable on certain rocks when they have been blasted during construction. The point of the cone points toward the applied impact. Figure 15.22 is an example of a Wells Creek shatter cone. Relative dating is currently the only means for dating the impact. The youngest rock that was apparently affected by the impact is late Mississippian in age (circa 326 Ma), and the oldest rock apparently unaffected by the event is the Tuscaloosa Gravel considered late Cretaceous in age (circa 100 Ma); thus the impact is believed to have occurred some time between 326 to 100 Ma.

The Flynn Creek Structure is located about five miles south of Gainesboro, Jackson County. The age of the structure is much older than the Wells Creek Structure and really does not belong in this section, entitled "The Last 250 Million Years," but because it is "genetically" related to a group of known impact structures in Tennessee, it is described here. Considerably smaller than Wells Creek, the structure is about two miles in diameter. The outcrops along Flynn Creek display many of the same types of elements associated with the Wells Creek Structure. Large breccia blocks with flowlike breccias filling some of the fractures, faults, and folds are some of the elements exposed. Two striking features of the Flynn Creek Structure are the uplifted dome at its center and the increased thickness of the Chattanooga Shale in the crater's depressed area. One is a rebound feature commonly associated with impact structures. In the center of the Flynn Creek structure is a sequence of folded, faulted, and brecciated rocks of the Knox Group that have been lifted about 1000 feet above their normal stratigraphic position to form a hill several hundred feet above the crater floor. The other striking element is a substantial change in the thickness of the Chattanooga Shale over a very short distance that delineates a basin formed after the impact. At the rim of the crater, exposures of the Chattanooga Shale are about 20 feet thick, but near the center, exposures are about 100 feet thick, suggesting that the mud forming the shale accumulated more thickly in the basin. As in the Wells Creek case, relative dating comes in to play. The Chattanooga Shale places the constraints on the age. The impact had to have occurred some time prior to deposition of the Chattanooga Shale, so it was likely some time during middle to late Devonian (398–385 Ma).

The Howell Structure near Fayetteville, Lincoln County, is smaller than the other Tennessee structures, and even though fewer features have been discovered, enough have been found to warrant recognition of the site as an impact structure. Brecciation, minor folding and faulting, and possibly two shatter cones have been noted (Deane and others 2004). The shape of the structure is roughly ellipsoidal and a mile and a half in diameter. Given a constraint placed again by the presence of undeformed Chattanoogan Shale, the impact may correlate with the Flynn Creek impact (398–385 Ma).

Cultural Geology

For many years, the northeast-southwest trending mountains of the Appalachians, including the Unaka or Great Smoky Mountains, the ridges of the Valley and Ridge province, and the steep escarpments of the Cumberland Plateau confined early American settlers to the coastal and piedmont regions east of the Blue Ridge Mountains (areas presently recognized as the Carolinas, Virginia, and Georgia). Even the adventuresome, once through the high mountains of the Blue Ridge via old Indian trails or through water gaps like those of the French Broad, the Nolichucky, the Little Tennessee, Pigeon, Hiawassee, and Ocoee Rivers, found themselves in a maze of ridges and valleys. However, many who braved the trek through the Blue Ridge were captured by the beauty of the region and settled in the isolated hollers, coves, and gulfs bounded by ridges or mountains. Other groups, often led by long-hunters or explorers such as Dr. Thomas Walker and Daniel Boone, sought still greener pastures and migrated along the Wilderness Road through the Cumberland Gap near Harrogate, Tennessee, (a wind gap in a ridge formed of deformed sandstone) into the Middlesboro Basin of Kentucky, an area that had been affected by a meteorite impact that occurred millions of years ago, and on farther westward to the interior lowlands.

Geology has profoundly influenced the socio-cultural, economic, and political development of every region on the Earth, and Tennessee is no exception. The settlers in the Tennessee Appalachians applied the names *gulf* and *cove* to many deep valleys or areas surrounded by uplands because on certain days when clouds hung low in them, especially when viewed from above, they resembled the water in coastal gulfs and coves—scenes very similar to the coastal homelands of Scots-Irish, Cornish, and other groups that made the decision to settle in the heart of the Appalachian Mountains. Coves, found mostly in the Unaka (Great Smoky) Mountains area, are, in terms of structural geology, called "windows," the result of erosion revealing the limestone bedrock footwall of thrust faults. These isolated coves, just like the isolated valleys (hollers), gorges, and gulfs in the Appalachians, often retain the names of the families who settled them; for example, Cades Cove, Wear Cove, Miller Cove, and Jones Cove. In the Valley and Ridge province, for example, there exist Hines Valley, Stanley Valley, and Powell Valley. Thus these isolated settlements were family oriented and quite autonomous. Each settlement was self-ruling, which sometimes resulted in conflicts or feuds between families in adjacent valleys or coves. The classic example is the famous Hatfield-McCoy feud of West Virginia/Kentucky.

The isolated hollers and coves cultivated a proud, strong, and self-reliant society that retained many of the manners of speech, music, and other customs that were brought to this country by the very first settlers (some of the old

Figure 15.22. A shatter cone collected from the Wells Creek impact structure. The orientation of the shatter cones (with the sharp end pointing upward) provide evidence that the force creating the fracture came from the impact of an extraterrestrial source.

customs prevail today). Handmade farm and home implements, along with musical instruments such as dulcimers, were crafted from the hardwoods that grew on the ridges. Other natural resources were soon exploited as especially the Cornish and Welsh descendants discovered deposits of iron. These immigrants were held in high esteem and were considered well versed in mining matters because they came from the mining areas of their homelands. Many little communities today retain names that refer back to their glory days of mining. For example, Ore Bank in Greene County, Red Bank in Hamilton County, and Lead Mine Bend in Union County.

16

Natural Resources

But as for us, though we may not have perfected the whole
art of discovery and preparation of metals, at least we can
be of great assistance to persons studious in its acquisition.

—Agricola, *De re metallica* (1550)

During Tennessee's rich mining history, many of its natural resources have been
mined and shipped around the world. Every county in the state has at one time
or another been a source for one or more of the nearly 40 mineral commodities
identified with economic potential (Upham 1992). Probably the first to exploit a
portion of Tennessee's mineral resources were the Native Americans who mined
flint and chert for tools and weapons, clay for pottery, salt for preserving food,
and galena and hematite/limonite for ceremonial body paint. Later the early
white settlers added limestone and clay for construction and sandstone for mill-
stones in the gristmills to the list of useful natural resources.

The branch of geology concerned with earth materials that can be extracted
for the well being of society is called economic geology. In general the three categories of mineral and rock resources included in the study of economic geology
are metals and metallic ores; mineral fuels; and industrial rocks and minerals (a
large group of materials, mostly nonmetallics). For the most part, these materials
are considered nonrenewable resources.

Ore is a term most often associated with mining, but before any mineral or
rock commodity can be considered an ore, or otherwise economically feasible
to mine, many factors must be considered. The most important of which are:
1) accessibility of the deposit; 2) hauling distances; 3) proximity to a water source
for mining purposes, or in some cases, cost of water removal; 4) market value of
commodity ("grade" or quality of the material); 5) distance to the market; 6) cost
of beneficiation (concentration of ore); 7) the reserves or abundance of the material; and 8) cost of reclamation. If a deposit does not favorably meet these requirements, it cannot be considered an ore. In some cases, however, a very high-grade
(or pure) ore deposit may occur with small reserves (quantity) and be considered

economical, or sometimes a low-grade deposit with vast reserves may be considered economical.

Metallic mineral deposits are typically the source for one or more metals such as zinc, gold, silver, lead, copper, mercury, and iron. Most of these deposits, however, do not occur as native metals. The metals normally occur as elements within a mineral along with other minerals forming a rock. The minerals in the deposit having no economic value are referred to as **gangue**. The gangue must be separated from the desired metal(s) in a milling process before the metal(s) can be sold at a profit. The **mineral fuels** category consists of major sources of energy, including the fossil fuels, coal, oil, and natural gas, as well as uranium minerals for developing nuclear energy. **Industrial rocks and minerals** are the category most commonly overlooked or taken for granted; however, in sheer volume and value, this resource category outranks all other categories. The construction industry alone consumes the highest volume of industrial rocks and minerals, including crushed stone, dimension stone (marble, limestone, and granite), and the raw materials for cement, brick, tile, and insulation. The chemical industry uses salt, sulfur, and limestone; agriculture relies on phosphate, nitrate, and potash; metallurgy requires graphite, fluorspar, magnesite, molding sand, and fire clay; the ceramics industry utilizes clay and feldspar. A host of other materials are used for abrasives, well-drilling mud, filtration media, fillers or additives, and lubricants.

Metallic Mineral Deposits

LEAD AND ZINC

Lead (the mineral galena, PbS) and zinc (the mineral sphalerite, ZnS) are sulfide minerals typically occurring in the same deposit. Native Americans first mined lead for ceremonial body paint, but by the early 1800s it was sought after to make bullets. The associated metal zinc finds its use in galvanizing iron (to prevent rusting) and as an alloy with copper in forming brass. Although the chief zinc ore mineral is sphalerite, the first zinc ore exploited in Tennessee included the minerals smithsonite ($ZnCO_3$) and calamine ($Zn_4Si_2O_7[OH]_2 \cdot H_2O$), both minerals formed by the weathering (oxidation) of sphalerite. In 1865 an open-pit operation was developed at Mossy Creek (now Jefferson City) where the smithsonite and calamine were extracted (Secrist 1924). Mining of sphalerite as the chief ore did not begin until about the time of World War I. Since then mining has been almost continuous in the Mascot–Jefferson City District. Other districts that have produced lead and/or zinc in East Tennessee include Bumpass Cove in Unicoi and Washington Counties, the Powell River District in Union County, and the Copper Ridge District in Grainger County. In the 1960s prospecting for zinc shifted to Middle Tennessee, and by the 1970s ore was being produced from the Gordonsville-Elmwood District in Smith County.

Zinc mineralization in most of the districts occurs in solution-collapse breccias in the upper part of the Knox Group, especially the Mascot Dolomite and the Kingsport Formation. The collapse breccias may very well be correlated with the post-Knox karst development at the close of the Sauk Sequence (the paleokarst unconformity between the Sauk and Tippecanoe Sequences). Sinkholes and solution channels in the Knox filled with blocks of broken rock served as conduits for the ore-bearing solutions that altered the bedrock and precipitated the ore minerals in the voids. Some of the typical gangue minerals often associated with the zinc deposits include pyrite, chalcopyrite, fluorite, barite, hematite, and dolomite. The Gordonsville-Elmwood District is particularly famous for beautiful fluorite crystals.

COPPER AND SULFURIC ACID

Copper was produced from the Ducktown-Copperhill District, Polk County, Tennessee, from about 1850 until about 1987, when economics forced the closure of operations. The earliest smelting of the ore by open roasting, using charcoal from

Figure 16.1. The Copper Basin (Ducktown), Tennessee, circa 1956. View is to the east across the depression caused by collapse of the Burra Burra Mine. The nearly 50 square miles of denuded land was caused by trees being cut for charcoal and used in early open roasting of ores and sulfurous gases produced from the roasting process. At the present time the area is almost totally reclaimed with vegetation.

surrounding trees, inflicted one of the earliest adverse environment impacts in the southeast. Besides decimating the forest, the sulfur dioxide fumes produced during the roasting combined with moisture in the atmosphere to produce acid rain (sulfuric acid). New vegetation was unable to replace the trees harvested to produce charcoal, and the land was ravaged by erosion (see figure 16.1). An area some fifty miles in diameter existed as a barren "badland" easily identified on satellite imagery. Reclamation in the last decades has been successful, and the barren wasteland is nearly covered with new vegetation. In 1907 the first sulfuric acid was recovered from the smelter effluent as a product, and eventually, until operations ceased in 1987, sulfuric acid became the chief product of the district.

The ore occurs in massive lenslike bodies within Neoproterozoic metamorphosed layers of graywacke and schist that have been folded and faulted. The bodies consist of a mixture of various sulfide minerals—mostly pyrite (FeS_2), the main ore chalcopyrite ($CuFeS_2$), and pyrrhotite (iron sulfide).

Underground mining at depths up to 3000 feet prevailed during most of the district's history until the last days of mining, when open-pit mining was attempted. The origin of sulfide minerals is thought to be associated with Neoproterozoic **black smokers**—chimneys rising from cracks on the ancient seafloor that spew hot mineralized water. The term *black smoker* arises from the cloud of tiny precipitating mineral grains that appear as smoke emanating from the chimneylike deposits.

Metals other than copper recovered in the milling process include iron, zinc, gold, and silver. It has been said that even though gold and silver would not constitute a feasible ore deposit, the revenue received from their sale defrayed much of the mine's operating costs.

IRON

Iron is no longer considered a mineral resource of Tennessee; however, through the years, three different minerals have been mined as iron ore—hematite (red iron ore), limonite (brown iron ore), and magnetite (magnetic or black iron ore). Mining iron in Tennessee dates back to Revolutionary War days, and it continued to thrive through the nineteenth and early twentieth centuries. Early iron production used furnaces built from blocks of limestone for smelting the ore into ingots called pig iron. Figure 16.2 shows the remains of a typical iron furnace used to make pig iron.

Some of the earliest iron mined was limonite, or "bog ore," which is found in thick residual soils overlying limestone and dolomitic rock units. Much of this early iron was mined and milled in northeastern Tennessee in places such as Embreeville in Bumpass Cove, Washington County. Charcoal was commonly used to roast the iron ore in these early mills and forges. Later, however, hematite (of-

Figure 16.2. This iron furnace constructed of quarried blocks of limestone in Houston County is typical of those used in the early iron production in Tennessee.

ten very limy and so considered a self-fluxing ore) was discovered as beds within the Silurian-aged rocks. A belt, including Whiteoak Mountain, that extends from Roane County through Rhea, Meigs, and Hamilton Counties is an example of where abundant hematite was once mined. Hematite also was mined along the boundary between the Cumberland Plateau and the Valley and Ridge provinces from Elk Valley, Campbell County, in the north to Lookout Valley, Hamilton County, in the south. These later deposits of iron were more favorable because coal was available from the rocks of the plateau, and limestone was available in rocks associated with the hematite—both materials needed to mill the ore. Iron furnaces cropped up in places like Rockwood (many iron mines in this vicinity), Dayton, and Chattanooga. Coke used in the production of iron was produced from coal in ovens at many locations along the eastern escarpment of Walden Ridge of the Cumberland Plateau. Dunlap, in Sequatchie Valley, has restored many of the old coke ovens that were operated there during the iron-producing days of the past (see figure 16.3).

Limonite, or brown ore, also known as bog ore, is a secondary mineral resulting from the alteration or solution of a previously existing iron mineral such as pyrite. "Limonite after pyrite" is a good example. It can be observed where limonite (amorphous or without form) replaces cubical crystals of pyrite. Many outcrops of Neoproterozoic rocks along the Ocoee River display perfect cubes of pyrite and limonite after pyrite. The main limonite deposits mined in Tennessee occur as large amorphous lumps within the residual soil blanketing the erosional surface on the Fort Payne Chert that caps the Western Highland Rim of Middle Tennessee. The principal counties that have produced brown iron ore include Stewart, Dickson, Lewis, Montgomery, Lawrence, Hardin, and Wayne. Lesser amounts of brown iron ore have been mined in East Tennessee, where the deposits occur within the residuum above various carbonate rock units in the Valley and Ridge province, such as the Shady Dolomite, the Knox Group, and the Rome Formation. Unicoi, Monroe, Carter, Washington, Cocke, McMinn, Greene, and Blount Counties were probably the most productive. Where massive sulfide mineral deposits, like those of the Ducktown-Copperhill District, have been exposed at the land surface, weathering (oxidation) forms what is known to prospectors as **gossan** (iron hat), a spongy-looking deposit of iron oxide plus other oxide minerals. Gossan has been mined as an ore of iron, but is more often considered an indicator of an underlying and potentially rich ore deposit consisting of the original ore mineral plus the minerals leached from the original rock. Ore deposits formed in this manner are known as **secondary enrichment**.

The main hematite deposits occur as sedimentary rocks in Silurian strata of Roane, Meigs, Hamilton, Campbell, Claiborne, and Marion Counties. The layers are typically very thin, but can range up to nearly seven feet in thickness. Wherever Silurian rocks occur in East Tennessee, it is not uncommon to find prospector's test pits dug into the hillsides. The hematite layers in the Rockwood Formation are calcareous (limy) and may be fossiliferous and/or **oölitic** (small spherical grains). Because limestone is important as a flux for removal of impurities during smelting, the carbonate content of the Rockwood hematite beds was thought to constitute a "self-fluxing" ore—not requiring limestone. Because the hematite beds are low grade and usually quite thin, only a few hematite mines were underground. Most of the mining was open pit. A few iron furnaces flourished along the base of the Eastern Escarpment of the Cumberland Plateau in towns such as Rockwood and Dayton until about the mid 1900s; then, basically, Tennessee was out of the iron and steel business.

Magnetite was mined intermittently from the late 1700s to the mid 1900s from lens-shaped bodies occurring along foliation in the Neoproterozoic Cranberry Gneiss in the vicinity of Roan Mountain, Carter County. An abandoned mine **adit** (mine opening) of a typical small magnetite mine in Tennessee has

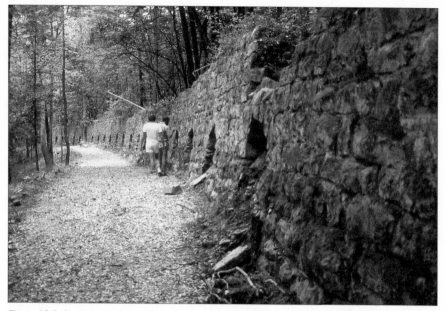

Figure 16.3. Restored coking ovens at Dunlap, Tennessee.

been preserved for viewing at Roan Mountain State Park (actually there are three abandoned mines on park property). However, the Tennessee mines were dwarfed by a larger magnetite mine with significant production just over the state line in the community of Cranberry, North Carolina. This mine operated sporadically until the later 1900s, producing a variety of related materials.

As higher-grade deposits with larger reserves were discovered outside of Tennessee, iron ore mining dwindled and finally became a thing of the past by 1960 (Upham 1992).

GOLD AND SILVER

Records indicate that gold was discovered at Coker Creek in southern Monroe County in 1827 (Ashley 1911; Troost 1837); however, Safford (1869) on the basis of records of the U.S. Mint in Philadelphia, reported the discovery date as 1831. Most of the mining occurred between 1838 and 1852, with a total production of 9000 ounces (Hale 1974). Although the native metal occurs in bedrock of the Ocoee Supergroup and in cross-cutting quartz veins, most of the gold has been recovered from **placer** deposits—sand and gravel of alluvium and colluvium formed from the weathered/eroded gold-bearing rock. As mentioned above, lesser amounts of gold and silver had been recovered during the milling of the sulfide ores of the Ducktown–Copper Hill District.

MANGANESE

Manganese is used variously as an alloy in metallurgy, an additive in gasoline, a pigment and a chemical reagent, and in dry-cell batteries. It was first mined on a small scale in Hickman County in 1837 (Reichert 1942). Psilomelane and pyrolusite, oxides of manganese and the principal ore minerals, occur as lumps within the residuum of carbonate rocks of East and Middle Tennessee. Manganese production from open pits peaked during the late 1940s and 50s and virtually ceased by 1960, when the U.S. government ceased its stockpiling program. Most of the production was from the residuum of dolomite of the Shady Dolomite and Rome Formation in Johnson, Carter, Washington, and Unicoi Counties.

ALUMINUM

Bauxite is the ore of aluminum. According to Floyd (1965) the entire bauxite production in Tennessee came from high-alumina clay residuum at seven mine sites—one in the Keenburg District of Carter County and six in the vicinity of Chattanooga in Hamilton County. Approximately 300,000 long tons of ore were produced between 1904 and 1926 (Floyd 1965). It is postulated that the bauxite formed in paleokarst features (ancient sinkholes) within the residuum of carbonate rocks such as the Shady Dolomite (Carter County) or the Knox Group (Chattanooga) weathered during the Tertiary Period. At present bauxite is not a commodity sought after in Tennessee.

TITANIUM, ZIRCONIUM, THORIUM, AND RARE EARTHS

These elements are typically found together. The titanium-bearing minerals ilmenite ($FeTiO_3$) and rutile (TiO_2) and related minerals containing zirconium, thorium, and rare earths (elements numbered 57 through 71 on the periodic table of elements), because they have physical and chemical properties making them resistant to weathering and erosion, are found in deposits of sand and gravel. Because of their high specific gravity, the minerals accumulate as distinct black layers with sand or sandstone called **heavy minerals**. Although heavy minerals can be found within almost any of the sandstone formations in Tennessee, the Cretaceous and Tertiary sands of the Mississippi Embayment hold the greatest potential for production of these commodities. At this time there is no known production in Tennessee.

Nonmetallic or Industrial Rocks and Minerals

LIMESTONE AND DOLOMITE

Limestone is a term that was first applied in a literal sense to mean a stone from which lime was produced; thus the origin of the word is tied to the mineral industry. However, in a modern geologic sense, the term *limestone* has come to refer to a large group of sedimentary rocks having a carbonate content, some of which would actually not be totally suitable for the extraction of lime.

Modern usage of the term *limestone* includes those sedimentary rocks composed of 50 percent or more of the minerals calcite ($CaCO_3$) and dolomite Ca Mg $(CO_3)_2$, in which the calcite is more abundant than dolomite. Generally, to be considered of commercial value, the combined carbonates must be 90 percent or more of the total rock. Examples of commercially used terms for carbonate rocks are the following:

>Dolomitic limestone—>10 percent mineral dolomite
>Magnesian limestone—>5–10 percent mineral dolomite
>High-calcium limestone—>95 percent calcite

The carbonate content as indicated in the examples above is the main criterion for commercial consideration; however, regarding the use of the carbonate rocks, there is no restriction based on origin, or other factors. Dolomite, a close relative to limestone, is a sedimentary rock with >50 percent calcite and dolomite, but in which the mineral dolomite is dominant. Some terms used for varieties of dolomite are the following:

>Calcitic dolomite—>10 percent calcite
>High-purity dolomite—>97 percent of total carbonate

In some cases the physical properties of limestone make it useful, and in other cases the chemical properties are more important. Limestone has a wide variety of uses; however, 50 percent of the limestone mined is in the form of crushed stone utilized as concrete aggregate, road metal, railroad ballast, rip rap, and, in finer sizes, sewage filter beds, French drains, poultry grit, stucco, manufactured sand, coal mine dust, whiting, and sulfur-dioxide flue scrubbers.

The early iron industry in Tennessee used limestone extensively as a flux to remove impurities during smelting and refining in the production of iron (see figure 16.4). As a soil amendment, the calcium in limestone is an essential nutrient for plant growth. Although fertilizers commonly refer to "agricultural lime" or "aglime," it is not true lime—it is typically only pulverized rock. Although calcium itself is an essential plant nutrient, it also plays an important role in making other nutrient elements such as magnesium, potassium, and iron available to plants through chemical reactions. True "lime" (CaO), a derivative of limestone, is a basic chemical of great importance in a long list of industrial applications, including glass making and rubber manufacture. Limestone as dimension stone is widely used in the construction of buildings, monuments, and a wide variety of statuary.

Figure 16.4. An abandoned limestone quarry near Dayton, Tennessee, where limestone was mined for use in an early iron industry.

Tennessee is blessed with abundant limestone resources. More than one-half of Tennessee's total area contains accessible carbonate rocks. High-purity carbonate rocks have been quarried in at least 60 of Tennessee's 95 counties (Hershey and others 1985). The Valley and Ridge province and Sequatchie Valley of East Tennessee, the Highland Rim and Central Basin provinces of Middle Tennessee, and the Western Valley of the Tennessee River are the principal locations of carbonate rock resources.

BARITE

Barite, barium sulfate, because of its high specific gravity (weight) has often been referred to as "white lead." When pulverized, the mineral dust is white, and legend has it that on occasion mill operators would add a little barite to their flour sacks to deceive farmers and retain a portion of the flour for themselves. Barite has been mined in many locations in Tennessee, but the major mining district is in Monroe and McMinn Counties in the vicinity of Sweetwater. Minable barite

Figure 16.5. A pinnacle of brecciated Knox dolomite in the Sweetwater barite district. The barite-bearing residuum has been mined from the pit surrounding the more resistant rock in the pinnacle.

typically occurs in residuum above the barite-bearing carbonate rocks such as limestone and dolomites of the Conasauga Group and the Knox Group.

Barite in the Sweetwater District, site of the most recently active mining operations, occurs in residuum derived from veins of barite and other minerals deposited within brecciated carbonate rocks similar to the lead- and zinc-bearing rocks in the Mascot–Jefferson City District referred to above (see figure 16.5). Many former barite mines and prospects are scattered throughout northeastern Tennessee, mostly in Greene, Washington, Carter, and Unicoi Counties. The barite in many of these occur within residuum from carbonate rocks in the Conasauga Group.

Barite is used in pharmaceuticals, various manufacturing industries, including glass. Also, because of its high specific gravity, it is used extensively in drilling mud.

PHOSPHATE

At one time Tennessee was an important producer of phosphate. Much like barite, phosphate is also mined from residuum. Residuum from the Ordovician formations, Bigby, Leipers, and Hermitage, particularly in Hickman, Maury, Giles, and Williamson Counties in south-central Tennessee, provided the main deposits. Columbia and Mt. Pleasant were the hubs for production and processing the ore. Depleted reserves have caused phosphate operations to cease in Tennessee.

Dimension Stone

TENNESSEE "MARBLE"

The East Tennessee region was at one time well known for its commercial "marble." It is not true marble in the geologic sense. It is perhaps better referred to as commercial marble because it is not a true metamorphic rock, but instead is a coarse, nonporous limestone that polishes to an attractive rock (see chapter 10). Tennessee marble, which is typically some shade of gray, brown, pink, or white, has been quarried principally in Knox and Loudon Counties; however, a dark-colored limestone is quarried near Thorn Hill in Grainger County and is marketed as Imperial Black marble. The marble industry in Tennessee dates back to 1838, when a quarry was opened near Rogersville in Hawkins County. Tennessee marble can be found in buildings in many states, including Florida, Georgia, Iowa, Massachusetts, Michigan, New York, Ohio, Pennsylvania, Texas, and Tennessee, but most prominently Tennessee marble can be found in some notable buildings in the U.S. capital—the National Museum, the National Art Gallery, the Taft Memorial, the Capitol Building, and the Lincoln Memorial. Although there is some current quarrying of Tennessee marble, its heyday has probably passed.

"CRAB ORCHARD" SANDSTONE

Certain Pennsylvanian-aged sandstone units on the Cumberland Plateau, especially in the vicinity of the community of Crab Orchard in Cumberland County,

Figure 16.6. Small Crab Orchard stone quarry along U.S. Highway 127, southeast of Crossville, Tennessee. Note the truck in the upper left corner for scale.

can be easily quarried into dimension stone used for building veneer, flagstone, and interior and exterior wall construction. The thin-bedded sandstone is readily separated, using wedges, into slabs or blocks ready for final sawing or breaking into desired shapes and sizes. Figure 16.6 is a typical small-scale Crab Orchard stone quarry along U.S. Highway 127 southeast of Crossville, Tennessee. Color bands of various shades of yellow, tan, gray, red, and brown, formed by oxidation of iron in flow patterns within the sand during the **lithification** process (converting sediment into a sedimentary rock), make the stone attractive. Buildings in the Homestead area near Crossville in Cumberland County offer some of the best examples of Crab Orchard stone. The Homestead community was one of the New Deal projects of President Franklin D. Roosevelt. The homesteaders cleared the land, quarried the rock, and built the homes, school, water tower, and other community buildings largely from Crab Orchard stone. Most of the stone quarries today are small family-type operations, but a few large producers still exist. Floyd (1965) reported that Crab Orchard stone has been shipped throughout the United States and even to some foreign countries.

Mineral Fuels

BITUMINOUS COAL

Bituminous coal has been mined in Tennessee for close to 200 years. Perhaps first used for domestic heating, the coal has been used in industry and now primarily for large-scale electric-power generation. The coal classified as medium- to high-volatile A-Bituminous is considered weakly to strongly coking and was used in the early iron/steel industry in Tennessee. As the metals industry in Tennessee has ceased, so has the coking process; however, some coking coal has been exported. Remnants of old coking can be observed at Dunlap, Tennessee (see figure 16.3).

Coal is found in rocks of the Absaroka Sequence of the Cumberland Plateau, the "Southern Tennessee Coal Field," and of the Cumberland Mountains, the "Northern Tennessee Coal Field." The coal seams in Tennessee are relatively thin (on the average about two to three feet in thickness) when compared to other coalfields in Appalachia. Bledsoe, Marion, Sequatchie, and Grundy Counties are the major producers in the Southern Coal Field, where surface mining is prevalent (see figure 16.7), and Claiborne and Campbell Counties are the leaders in the Northern Field, where underground mining is more common (see figure 16.8).

LIGNITE

Lignite, a dark-brown, woody deposit, is actually a stage after peat in the formation of bituminous coal and is considered by some to be a low-grade coal. The

Figure 16.7. Surface mining in Sequatchie County, Tennessee, where surface mining is more prevalent than underground mining.

lignite occurs in deltaic deposits of Cretaceous and Eocene ages in the Mississippi Embayment. The thickest deposits, ranging from less than a foot to nearly 10 feet, occur in Dyer, Crockett, Haywood, Lauderdale, and Tipton Counties (Luppens 1980). Despite speculation that there are very large reserves of the low-quality lignite in Tennessee, demand has not warranted mining the resource.

URANIUM

Uranium prospecting was quite active in Tennessee during the 1950s. Radioactivity in the Neoproterozoic crystalline basement rocks (Cranberry and Beech) in the Roan Mountain region prompted considerable activity during this period. However, no ore-grade concentrations were discovered. It is unlikely that these rocks will be mined in the foreseeable future.

The Kaskaskian Chattanooga Shale represents a deposit of low-grade uranium with significant reserves. The Chattanooga underlies all of the Cumberland Plateau and the Highland Rim and parts of the Central Basin. Economic conditions would have to be just right to justify the deep underground mining that would be required. However, the Chattanooga Shale is a more probable resource than the Neoproterozoic rocks in northeastern Tennessee.

PETROLEUM

The term *petroleum* refers to a naturally occurring mixture of hydrocarbons (compounds of carbon and hydrogen) and various amounts of nitrogen, oxygen, and sulfurous impurities. When the mixture is distilled, it yields a wide assortment of fuels and petrochemicals, including pesticides and herbicides. Hydrocarbons ordinarily consist of a liquid phase, called crude oil, and a natural gas phase, which is mainly methane (CH_4). Of the two hydrocarbon phases, natural-gas production probably plays a larger role in Tennessee's economy.

Morgan, Scott, Claiborne, and Fentress Counties have been the major producing areas. The hydrocarbons occur primarily in **stratigraphic traps** (reservoir rocks) within the Ft. Payne, Monteagle, or Bangor Formations—where the underlying carbonaceous Chattanooga Shale is considered to be the likely source rock. Other rock units that have some potential for production include the Knox Group, the Trenton (Nashville) Group, and the Stones River Group (Zurawski 1981).

OIL SHALE

The carbonaceous Chattanooga Shale, because of its dark organic color and extensive occurrence, as noted above, has been studied as a potential source of hydrocarbons through distillation. Some local folks have actually burned the shale as fuel in their stoves. Although it has been distilled to form a thick tarry fluid, unless a severe energy crisis occurs, the Chattanooga will not serve as a fuel source. However, it is considered the source rock for hydrocarbons trapped in the overlying reservoir rocks in the areas mentioned earlier where petroleum is produced.

Gems

DIAMONDS

An unusual igneous pluton in Union County was first recognized by James Safford and reported in his 1869 seminal work, *The Geology of Tennessee.* This site today lies on the Central Peninsula of Norris Lake across the lake from the Hickory Star boat dock. In 1904 two diamonds were discovered in gravel of the Clinch River (now Norris Lake) by fresh-water pearl fishermen (Hall and Amick 1944). The diamonds are attributed to the igneous pluton whose composition, peridotite (see chapter 3), is typical of diamond-producing rocks around the world.

Magnetite and garnet crystals still can be found in the part of the pluton exposed above the water of Norris Lake. The pluton has been altered significantly to a weathered chlorite schist by subsequent metamorphism and present-day weathering. As reported by Hall and Amick (1944) one diamond in the rough weighed about three carats, and the other weighed about seven-eighths of a carat. The larger diamond was cut to a stone of one and one-fourth carats, and the smaller to a stone of three-fourths of a carat. To this day, no other diamonds have been found.

Interestingly, the Tennessee River pearl that was being sought when the diamonds were discovered was designated as the Tennessee State Gem in 1979. This is ironic since the river pearl, being organically formed, is not a true mineral. The pearls are created by fresh-water mussels (bivalve organisms similar to oysters) and can be found in all colors, shapes, and sizes. They are especially prized for their great durability. Tennessee is the only state that has a "biomineral" for a state gem.

Afterword

Perhaps one of the most difficult aspects of writing this book was paring the voluminous record of Tennessee's geologic history into "Cliffs Notes" form while presenting the pertinent facts in a clear, chronological, and understandable fashion. If the book has been successful in that respect, the reader now should have an appreciation for the geologic ideology underpinning it:

1. Tennessee has a rich geologic history that is in no way stagnant. As new theories about Earth processes evolve and new technologies allowing more perceptive observations of the rock record are developed, the story continues to become more refined and complete. For example, just a few decades back, anyone who even contemplated a theory such as plate tectonics was considered a heretic. Now, just think how GPS and other advances in science have facilitated mapping, not only of our own planet, but other bodies in the universe. Also, our observational senses have been extended into the atomic realm by instruments and analytical techniques such as the electron microprobe and isotopic analyses, respectively.

2. Earth's systems interact every day on a wide variety of scales. Ranging from small grains of sand moving along networks of rivers and streams requiring millennia to reach the ocean to, at the same time, large earth movements. Tsunamis deposit tons of sediment in a matter of seconds, and although Earth's plates move very slowly, they can move suddenly causing earthquakes..

3. The Earth is very old. Sudden changes such as tsunamis, earthquakes, and volcanoes are mere spikes on the chart of the long-term changes that have profoundly shaped the Earth for more than 4.5 billion years. The fact that Earth history embraces billions of years (deep time) makes biological and landscape evolution easier to comprehend. Mountains have been built and then eroded away over the course of Tennessee's geologic history. Change is the one constant in Earth history.

4. Earth's systems have always been interacting, as can be discerned, especially from the fossil record. The atmosphere, hydrosphere, geosphere, and cryosphere of ancient environments (paleoenvironments) influenced the ancient biosphere (ecology/paleoecology). Likewise, the present geologic setting has had an impact on the human ecology of Tennessee. The mountains, valleys, lowlands, and uplands have molded distinct cultural heritages across the state of Tennessee.

5. There is a history of geology. There has been evolution of thought concerning interpretation of Tennessee's geology, and the thinking keeps evolving with new theories and new technology. This book will certainly have an addendum as cadres of different geologists and new technologies revise the story as told in this book or certainly add to it.

6. "The present is the key to the past." Always keep in mind James Hutton's basic principle and its corollary: the past should serve as a key to the future. An understanding of that corollary could mean the future survival of civilization as we know it today.

Two companion books are recommended for this book: Harry Moore's *A Geologic Trip across Tennessee by Interstate 40* (1994) and a forthcoming book on the fossils of Tennessee by Michael Gibson and Michael McKinney (in preparation).

Glossary

absolute age dating The determination, usually by use of radioactive materials, of the actual number of years before the present that a rock, fossil, or other artifact formed.

absolute time A specific date in years applied to geologic events as compared to relative time, where events are dated from before or after a given datum point.

abyssal plain The deep, flat portion of the ocean floor located some distance from the shallower **continental shelf** and **mid-ocean ridge** regions.

accreted terrane A block of crust containing igneous, metamorphic, and **sedimentary rock** units that is bounded by faults and is distinctly different from surrounding crustal blocks.

active continental margin Where the transition from thick continental crust to thin oceanic crust is at or near a current plate boundary and is thus a region of tectonic activity such as earthquakes, volcanic eruptions, and mountain building.

adit A horizontal portal into a mine on a hillside.

aggradation An external process of change whereby landforms are constructed by the accumulation of earth materials. Features of aggradation include many glacial deposits, river **deltas**, **barrier beaches**, **natural levees**, etc.

Alleghanian orogeny The convergent orogenic event that occurred about 270 Ma (million years ago) when Africa collided with North America.

alluvium Sediment that has been deposited by a river or stream.

amygdaloidal basalt Lava flow with almond-shaped gas pores that are often filled with minerals such as native copper, chlorite, fluorite, etc.

andesite A fine-grained, light-to-dark colored **igneous rock** that is about 60 percent silica.

angular unconformity An **unconformity** in which the underlying rocks are inclined at a different angle than the overlying strata.

anhedral grains Crystalline mineral grains without well-formed crystal faces. Typical of detritus.

antecedent stream A stream that cuts across an uplifted region (mountain or ridge); the stream must have existed before the region was uplifted and must then have been able to downcut as fast as the region was being uplifted.

anticline A **fold** in which rock layers are bent upward.

aphanitic A textural term used for igneous rocks, meaning that mineral grains are not visible to the unaided eye.

Appalachian Mountains An ancient highland region extending from the Atlantic provinces of Canada to Alabama, formed during ocean basin closure and continental collision during the **Paleozoic Era**.

arch 1. A broad, circular, upward **fold** in layered rocks (**dome**). 2. A curved erosional feature resembling a bridge where a hard rock layer is eroded completely through, leaving a continuous roof intact.

arkose A **sedimentary rock** containing a high percentage of feldspar minerals.

asthenosphere The relatively soft and plasticlike portion of Earth's upper **mantle**. The rigid plates of lithosphere drift about, gliding over the asthenosphere.

atmosphere The gases enveloping Earth.

banded gneiss A metamorphic rock formed by high temperature and pressure conditions that cause minerals to flow into nearly parallel bands of light- and dark-colored **minerals**.

barrier island An elongated bar of sand and mud above sea level that formed offshore and parallel to the coastline of the mainland.

basalt A fine-grained, dark-colored **igneous** rock consisting mainly of minerals rich in iron and magnesium and about 50 percent **silica**.

basin 1. A broad, circular, downward fold in rock layers (circular **syncline**). 2. A depression that accumulates sedimentary deposits.

batholith A very large igneous intrusion typically composed of **granite** classified as a **discordant pluton**.

bioclastic A mechanically deposited sediment composed of particles (**clasts**) from the hard parts (i.e., shells) of organisms.

bioherm A buildup by marine organisms such as a reef.

biosphere All living matter on Earth.

black smoker An emanation of sulfurous gases from the seafloor. Possibly related to "hot spots" in the mantle.

breccia 1. A **sedimentary rock** composed of angular fragments (**clasts**). A similar rock with rounded clasts would be classified as a **conglomerate**. 2. A zone of angular rock fragments resulting from grinding along a

fault. 3. Angular rock debris formed in collapsed **sinkholes**.

bridge As in natural bridge—an **arch** with a stream flowing underneath.

calcite A **mineral** composed of the compound calcium carbonate ($CaCO_3$). The primary constituent of the rock **limestone** and its metamorphic equivalent **marble**.

carbonate rock Rocks such as **limestone** or **dolomite** where the composition is largely (CO_3)-2.

cave (cavern) A void in the Earth's crust caused by the dissolution of rocks such as **limestone**, **dolomite**, or **salt**.

Cenozoic Era That portion of geologic time from about 65 Ma (million years ago) to the present. The term means "age of recent life."

chert A **sedimentary rock** formed of microscopic crystals of the mineral **quartz.** It may occur as irregularly shaped nodules or layers, or as the chemical replacement of organic remains such as **cyanobacteria**. Layered chert is formed either by precipitation from seawater or through the accumulation on the seafloor of dead single-celled organisms with siliceous shells. Most chert is buff to gray in color, but other varieties may be black (flint) or red (jasperoid).

clast A fragment broken from a preexisting rock that is mechanically transported and deposited as sediment.

clastic sedimentary rock A **rock** formed from the eroded debris from other rocks.

cleavage 1. A preferred direction for minerals to break based on the mineral's **crystal** structure. 2. Parallel planar fractures in a rock induced by rock deformation.

colluvium Rock debris on slopes transported primarily by gravity with the aid of water.

compression When diametrically opposed forces act against one another in a squeezing fashion. The action may cause a substance to fracture or decrease in volume.

conchoidal fracture A smoothly curving surface along which materials with no **cleavage** planes tend to break.

concordant Where the contact between a pluton and the intruded rock is parallel to the layering of the intruded rock. A sill is a concordant pluton.

conglomerate A **sedimentary rock** formed from compaction and cementation of sediment containing pebbles and cobbles (if the particles are angular, the rock is a **breccia**).

conodont A very small fossil jawbone of a wormlike organism.

contact The boundary surface between two rock bodies (as between two rock **formations**, between an **igneous intrusion** and adjacent rock, between two igneous rock bodies, or between rocks juxtaposed by a **fault**.

continental collision zone A mountainous region formed where an ocean basin closes and blocks of **continental crust** collide.

continental craton See **craton**.

continental glacier A vast ice sheet that spreads over thousands of square kilometers.

continental platform A part of the **continental craton** that is periodically invaded by shallow seas and becomes covered with layers of sediment.

continental rift zone An elongate mountainous region with long valleys formed when diverging plates pull apart a plate capped with thick continental crust.

continental rise The gently sloping part of an ocean basin between the deep **abyssal plain** and the steep **continental slope**.

continental shelf The shallow, flat region of an ocean basin immediately adjacent to a continent.

continental shield The region of the **continental craton** consisting of old igneous and metamorphic rocks exposed at the land surface.

continental slope A relatively steep portion of an ocean basin between the shallow water of the **continental shelf** and the deeper water of the **continental rise**.

convection A transfer of heat where warmer, less dense material rises while cooler, more dense material sinks. As in convection currents within Earth's **mantle**.

convergent plate boundary A region where two slabs (plates) of Earth's outer shell (**lithosphere**) move toward one another while destroying lithosphere. **Subduction zones** and **continental collision** are manifestations of convergent plate boundaries.

coprolite The undigested portion of an organism that has passed through the alimentary canal of another organism—fossil feces.

coral reef A buildup of the remains of the hard parts of coral and other marine organisms.

core The central part of Earth consisting mostly of iron.

correlation A process whereby the mutual relationships, i.e., age and environment, are established between strata in one locality with the strata in another locality.

craton A long-lived block of durable continental crust commonly found in the tectonically stable interior of a continent.

cross-bedding A pattern of inclined layers of sedimentary layers formed by water or wind currents shifting directions during deposition of sediment. Commonly seen in sand dunes developed on a beach or in a desert.

cross-section A diagram showing the configuration of earth materials underground as they would appear on an imaginary vertical slice through the Earth.

crust The outermost part of Earth composed mostly of light **silicate minerals**.

cryosphere Includes all ice on Earth's surface.

crystal A single, continuous piece of a mineral bounded by flat surfaces reflecting the internal molecular structure of the mineral as the mineral naturally grew.

crystalline Possessing an orderly internal molecular structure.

cyanobacteria Blue-green algae. A type of archaebacteria.

daughter isotope The decay product of radioactive decay.

décollement The lowermost glide zone in a thrust-faulted region. This zone in Tennessee is the Rome Formation—it is the first incompetent rock unit above the crystalline basement rocks.

deep-sea trench See **trench**.

deep time The history of Earth from its origin to the present measured in millions and billions of years.

deflation The process of erosion by which sediment is removed by wind.

deformation A change in shape, position, or orientation of a material by bending, breaking, or flowing.

degradation The external process of change whereby landforms are created by **erosion** by wind, water, ice, gravity, or organisms. Includes features such as **caves, sinkholes,** canyons, **natural bridges, and arches**.

delta A wedge of sediment deposited at the mouth of a river when the flowing water loses velocity on entering a standing body of water. The loss of velocity reduces the water's competence to carry sediment, and thus the sediment is deposited.

delta plain The low, swampy land on the surface of a **delta**.

dendritic drainage A drainage pattern where interconnecting streams resemble the branches connecting to a deciduous tree.

density Mass per unit volume.

deposition The accumulation of sedimentary material transported by wind, water, ice, wind, or gravity or by chemical or biochemical processes.

differential weathering/erosion Where the rates of **weathering** or **erosion** of earth materials vary because of differences in composition, orientation of exposure, or the intensity of the weathering/erosion agent.

dike A tabular-shaped discordant igneous **intrusive rock** (**pluton**) whose boundaries are not parallel with the layering of the intruded rock.

disconformity An **unconformity** where the overlying layers of rock are parallel to the layering of the underlying rock.

discordant When the contact between a pluton and the intruded rock is not parallel to the layering of the intruded rock. See **dike**.

dissolution A process during which materials dissolve in water.

distributaries The streams that fan outward as a delta is built.

divergent plate boundary A region on Earth where two slabs (**plates**) of the outer shell (**lithosphere**) are separating from another, creating new lithosphere.

divide The higher ground that separates watersheds/drainage basins.

dolomite A **carbonate sedimentary rock** similar to **limestone** but with some of the calcium in the carbonate compound replaced by magnesium [Ca, Mg(CO$_3$)].

dolomitization The geochemical process whereby limestone is **converted to dolomite**.

dome A broad, circular upward **fold** in rock layers (a circular **anticline**).

downcutting A process whereby water flowing through a channel cuts into the substrate and deepens the channel relative to its surroundings.

drag fold A **fold** that develops in rock layers adjacent to a **fault** during or just before slip.

dripstone Limestone (travertine in a cave) formed by the precipitation of **calcite** (calcium carbonate) from percolating groundwater.

drop stone A rock fragment dropped into a sedimentary layer from a floating raft of ice when the ice ultimately melts.

ductile When material can be stretched or drawn. Ductile material flows (like Silly Putty or hot plastic).

duplex faults Where thrust faults have ramped up on one another causing duplication of rock units.

dynamic metamorphism Metamorphism wrought as a consequence of shearing alone, with no change in temperature.

dynamothermal metamorphism Metamorphism that includes heat, pressure, and shearing.

earthquake A sudden movement within the Earth that releases vibrations.

earth system The global interconnecting web of physical and biological phenomena, involving the interaction of the **atmosphere**, **hydrosphere**, **cryosphere**, **geosphere**, and **biosphere**.

ecosystem A community of organisms and their physical environment considered as a unit.

embayment A low area of coastal land.

eon A major geologic time unit that embraces smaller units called **eras**.

eolian Sediment transported by wind.

epeirogenic movement A gradual uplifting or subsidence of a broad region of the Earth's surface, such as a continent.

epeirogeny An event of epeirogenic movement. The term generally refers to the broad (continental in size) movements of Earth's crust including the formation of mid-continent **domes** and **basins**.

ephemeral (intermittent) stream A stream whose bed lies above the water table, so that the stream flows only when the rate at which water enters the stream from rainfall or melt water exceeds the rate at which water infiltrates the ground below.

epicontinental sea A shallow sea overlying a continent.

epoch A geologic time unit that is a division of a **period**.

era A geologic time unit that is a division of an **eon**.

erosion The breakdown and removal of weathered rock debris by mechanical action of wind, water, ice, or gravity or by biological and/or chemical activity.

erratic A boulder whose composition is foreign to the bedrock on which it rests. Applied particularly to boulders that have been transported by **glaciers**.

escarpment A steep declivity that bounds an upland area such as a **plateau** as it is gradually eroded.

euxinic An environment with little or no oxygen.

evaporite A mineral deposit such as salt that has been deposited by evaporation from seawater.

extension Where the Earth's forces pull rocks apart, resulting in **normal or gravity faults**.

external process A change to the Earth related to erosion and deposition by wind, water, ice, gravity, and organic activity.

extrusive rock An **igneous rock** that solidified from **magma** that erupted on the Earth's surface (**lava**), to form fine-grained mineral crystals (**volcanic rock**).

fault A fracture in the Earth along which blocks of rock on either side of the fracture move parallel to the fracture plane.

fault plane The surface of a **fault** along which blocks of rock have been displaced.

fault trace (or line) The intersection between a **fault** and the ground surface.

feldspar One of a group of alumino-silicate minerals common to rocks in the Earth's crust.

floodplain The flat land on either side of a stream or river that becomes covered with water during a flood.

flow fold A fold the forms when the Earth's material is so soft that it behaves as a plastic.

fluvial deposit Sediment deposited by a stream or river within its channel or on its **floodplain**.

fold A bend or wrinkle of rock layers or foliation as a result of rock deformation. **Anticlines** and **synclines** are typical examples.

foliation Layering formed by the alignment of mineral grains or of compositional banding in a **metamorphic** rock.

foot wall The block of rock that lies below a **fault plane**.

foreland The region in front of a mountain range, sometimes the site of deformation as a **foreland fold-and-thrust belt** and sedimentary deposition in a **foreland basin**.

foreland basin The depression in front of a rising mountain range where sedimentary material accumulates.

foreland fold-and-thrust belt The deformed region in front of the main region of mountain building, where sedimentary layers are folded and transported along **thrust faults**.

formation A fundamental mappable rock unit defined by well-defined boundaries(at a map scale of 1:24,000). Typically named for the type of rock and geographic location where it is well exposed.

fossil The remains or trace of an ancient living organism preserved in rock or sediment.

Ga Abbreviation for a billion years.

gabbro A coarse-grained, dark-colored intrusive igneous rock with about 50 percent silica and rich in iron and magnesium minerals. The equivalent of **basalt**.

gangue Noneconomic minerals associated with an ore deposit.

geologic column A composite stratigraphic chart that represents the sequence of rock units within a region.

geologic history The sequence of geologic events that have taken place within a region.

geologic map A map representing the distribution of rock units and structures within a region.

geologic time The span of time from the Earth's origin to the present.

geologic time scale A scale describing the intervals of **geologic time**.

geology The study of the Earth, including its composition, behavior, and history.

glacial outwash The sediment deposited by melt water flowing from the terminus of a **glacier**.

glacier A long-lived river or sheet of ice that slowly flows over the land surface.

glassy A solid in which the atoms are not arranged in an orderly pattern.

glauconite An iron potassium clay mineral characterized by its green color. Formed through biochemical alteration of biotite mica during conversion of sediment to **sedimentary rock**. The process is often influenced by decay of organic matter. Found mostly associated with continental shelf sediments.

gneiss A **metamorphic** rock with bands of light- and dark-colored minerals.

Gondwana A **supercontinent** that comprised modern South America, Africa, Antarctica, India, and Australia. Also called Gondwanaland.

gossan Literally means "iron hat." A deposit of iron-oxide minerals (mainly hematite) occurring above ore deposits. The principal ore minerals have been leached from the surface and are concentrated at depth as a secondarily enriched ore deposit.

graben A down-dropped crustal block bounded on either side by a **normal fault** inclined toward the **basin**.

gradation External processes of change involving **erosion** and **deposition**.

grain A fragment of a mineral **crystal** or group of crystals. Usually found as a component of a rock.

granite A coarse-grained, generally light-colored igneous rock with about 70 percent silica.

graptolite An animal belonging to the phylum hemichordata. Considered index fossils for the Ordovician Period. Typically preserved as a carbonaceous film.

graywacke A term used for a **sedimentary rock** consisting of sand-sized up to small pebble-sized **grains** of **quartz**, **feldspar**, various **ferromagnesian silicate** minerals, and rock fragments all mixed together in a muddy matrix. Typical of **graded bedding** in **turbidite** deposits.

greenschist facies The lowest **metamorphic grade**, in which the mineral chlorite has formed.

greenstone A low-grade metamorphic rock formed from basalt; if foliated, the rock is called **greenschist**.

Grenville orogeny An **orogeny** that occurred about 1 billion years ago and yielded the belt of deformed and metamorphosed rocks that underlie the eastern fifth of North America.

group A succession of rock units (formations) lumped together on the basis of some commonality.

Hadean A geologic time unit. The oldest **Precambrian eon**, representing time from the origin of the Earth and the first rocks preserved.

half-life The time it takes for half of a group of radioactive element's **isotopes** to decay.

halite Sodium chloride, salt. A typical mineral found in **evaporite** deposits.

hanging wall The block of rock that lies above an inclined **fault plane**.

hardness A mineral's resistance to abrasion.

headward erosion The process whereby a stream channel is lengthened as the stream erodes upwards towards a **divide**.

heavy minerals Mainly insoluble iron and titanium minerals (dark in color and high specific gravity) found in mechanically deposited sediment.

hiatus A gap in the geologic record caused by the removal of the sedimentary record by erosion or by nondeposition over a span of time; as in the interval of time between the deposition of the youngest rock below an **unconformity** and deposition of the oldest rock above the unconformity.

high-grade metamorphic rocks Rocks metamorphosed under relatively high temperature.

Holocene The **period** of geologic time since the last continental glaciation.

horst The high block between two **grabens**.

hot spot A location at the base of the **lithosphere**, at the top of a **mantle plume**, where the temperature can cause melting.

hot spot track A chain of volcanic features created as a plate migrates over a **hot spot**.

hummocky topography An irregular and lumpy ground surface.

hydrosphere A subsystem within the **Earth system** that includes the Earth's surface water (lakes, rivers, and oceans), groundwater, and liquid water in the **atmosphere**.

hydrothermal deposit A deposit of **ore** minerals precipitated from hot-water solutions circulating through a **magma** or through rocks surrounding an **igneous intrusion**.

ice age An interval of time when the climate was cold enough to support ice sheets (**glaciers**) that periodically advanced over large areas of continents and many mountains supported glaciers.

igneous rock A rock that forms from the solidification of molten rock (**magma** or **lava**).

incised meander A river **meander** that lies at the bottom of a steep-sided canyon.

index minerals Minerals that serve as indicators of **metamorphic grade**.

inner core The inner section of the Earth's core consisting of solid iron alloy and extending from about 3200 miles (5100 kilometers) depth to the Earth's center at 4000 miles (6300 kilometers) depth.

internal process A change to the Earth related to heat energy within the Earth, including plate motion, mountain building, and/or **volcanism**.

intertidal zone The coastal area across which tides rise and fall.

intrusive contact The boundary between a **pluton** and the invaded rock.

intrusive igneous rock A body of rock (**pluton**) formed by the solidification of **magma** below the land surface.

island arc A curved chain of **volcanoes** formed where a plate capped by oceanic crust is **subducted** beneath another plate capped by oceanic crust.

isograd A line of equal pressure-temperature, based on **index minerals** drawn on a map to represent metamorphic grade.

isopach map A map with contours lines showing the thickness of a given rock unit.

isostasy A condition whereby a buoyant force pushing earth materials upward is balanced by an equal gravitational force (weight) pushing downward.

isotopes Different species of a given element that have the same atomic number but different atomic weights.

joint A fracture in a rock with no significant lateral movement of the blocks of rocks on either side of the fracture. Parallel joints form a set, and more than one set constitutes a joint system.

karst A region whose topography is caused by dissolution of limestone bedrock. Caves, **sinkholes,** and spires of limestone are typical features.

klippe Where erosion has left a remnant of the hanging wall of a thrust fault bounded by the fault trace and younger rocks of the foot wall.

K-T boundary event The mass extinction of organisms at the close of the Cretaceous Period, 65 million years ago, and before the Tertiary Period, possibly a result of the collision of multiple bolides with Earth. Aalso be referred to as the K-P boundary (Cretaceous-Paleogene).

lacustrine A lake environment.

lapilli Walnut-sized volcanic ejecta.

Laurentia A continent in the early **Paleozoic Era** composed of modern North America and Greenland.

lava The surficial equivalent of **magma**. The term is used for both the flowing liquid and the solidified rock. Lava can flow from **volcanoes** or from fissures in the land surface.

limb The side or flank of a fold (**anticline** or **syncline**).

limestone A **carbonate sedimentary rock** composed of calcium carbonate. **Grains** of the mineral **calcite** may accumulate mechanically, chemically, or biologically.

lithification The process whereby sediment is converted into **sedimentary rock** by compaction and cementation.

lithosphere The rigid outer shell of the Earth composed of the outermost **mantle** and

crust. The lithosphere is broken into plates that move over the underlying, softer mantle or **asthenosphere**.

löess Fine-grained sediment deposited by wind. Large loess deposits have formed from **deflated,** fine-grained glacial **outwash sediment.**

low-grade metamorphic rocks Rocks resulting from low temperatures.

luster The way a mineral looks in reflected light.

Ma Abbreviation for a million years.

mafic A term used to refer to a **magma** or **igneous** rock that is relatively poor in silica and rich in iron and magnesium.

magma Molten rock beneath the Earth's surface.

mantle The portion of the Earth between the **crust** and **core** made of **silicate minerals** rich in iron and magnesium.

marble 1. A **metamorphic** rock formed from **limestone** or **dolomite**; 2. **Crystalline** limestone that polishes well and can be sold commercially as marble.

mass movement (or **mass wasting**) The transportation of rock or regolith by gravity.

meander A snakelike bend in a river's course.

Mesozoic Era The age of middle life; the portion of geologic time between 248 and 65 million years ago.

metamorphic facies A **metamorphic** rock with a distinct suite of minerals formed in a specific environment of temperature and pressure.

metamorphic rock A pre-existing rock (**igneous**, **sedimentary**, or **metamorphic**) that has been subjected to a higher temperature and/or pressure than the conditions under which it was originally formed.

metamorphic zone The region between two metamorphic **isograds**, typically named after an **index mineral** found within the region.

metasedimentary rock A **sedimentary rock** that has been metamorphosed, but not to the extent that its sedimentary characteristics have been destroyed. Includes rocks such as metasandstone, metasiltstone, metagraywacke, and metaconglomerate.

microcontinent A small block of continental **crust** that broke off a larger continent.

mid-ocean ridge An undersea mountain range formed from volcanic activity where lithospheric plates diverge.

mineral A naturally occurring inorganic solid compound with a characteristic internal atomic structure with physical and chemical properties that are fixed or vary within a given range.

Mississippi Valley–type (MVT) ore An ore deposit, typically in dolomite, containing lead- and zinc-bearing minerals.

Mohs hardness scale Ten minerals ranked in order of relative resistance to abrasion used in using physical properties to identify minerals.

natural levees Low ridges on either side of a river or stream formed by sediment deposited during flooding.

Neoproterozoic The youngest era of the Proterozoic Eon (1000 Ma to 542 Ma).

nonclastic sedimentary rock A **sedimentary rock** formed by chemical or biochemical precipitation of material from a solution

nonconformity An **unconformity** where **sedimentary rock**s rest on an erosional surface on **igneous** or **metamorphic** rocks.

nonfoliated metamorphic rock A rock consisting of minerals recrystallized during **metamorphism**, but without **foliation**.

nonmetallic mineral resources Economic mineral resources that do not contain metals. Also referred to as industrial mineral and rock resources; examples include limestone, gypsum, dimension stone or building stone, sand, phosphate, and salt.

normal fault A fault where the **hanging wall** has apparently moved down relative to the **foot wall**.

obsidian An **igneous** rock with glassy **texture** that cooled so rapidly from magma (**lava**) that **crystals** did not have sufficient time to form.

oceanic crust The outer layer of Earth that is relatively thin and low in silica compared to **continental crust**.

ophiolite A sequence of rocks composed of oceanic crust and the uppermost mantle. From bottom to top, the sequence consists of the igneous rocks peridotite, gabbro, and basalt, overlain by a deep-ocean **sedimentary rock**.

oölite A spherical body, 0.25–2.00 mm. in diameter. Usually calcaeous, but may be siliceous, hematitic, or other composition. Often made of concentric layers precipitated around a nucleus such as a sand grain.

oölitic A **sedimentary rock** texture consisting of oölites.

orogeny A geologic event that forms mountains.

Ouachita Mountains A mountainous region in western Arkansas and eastern Oklahoma, formed by the same ocean basin closure and continental collision that formed the Appalachian Mountains.

outer core The zone of dense, liquid iron and nickel extending from 1800 miles (2900 kilometers) to 3200 miles (5100 kilometers) depth between the solid **lower mantle** and the solid **inner core**.

oxbow lake A crescent-shaped lake formed when a meander of a meandering river or stream has been cut off from the river or stream.

paleoclimatology The study of climates during the geologic past.

paleogeology The geology (topography, bedrock, and structures) as they would have been at any point in the geologic past.

paleogeologic map A map that portrays the topography, bedrock, and structures as they would have been at any point in the geologic past.

paleosol Ancient soil preserved in the **sedimentary rock** record.

Paleozoic Era The geologic time between 543 and 248 million years ago. The term *Paleozoic* means ancient life.

passive margin A continental edge that is not a plate boundary.

Pangaea A supercontinent that assembled at the end of the **Paleozoic Era**.

parent isotope A radioactive isotope that undergoes decay.

pegmatite A very coarsely textured **igneous** rock usually in the form of a **dike**. **Crystals** may be tens of centimeters up to meters in diameter.

peneplain A land surface eroded to nearly a flat surface near to sea level; **correlative** terraces and ridge crests are thought to be the elevated remnants of long periods of erosion.

peridotite A coarse-grained, olive-green igneous rock with about 40 percent silica that makes up most of the Earth's **mantle**.

period A geologic time unit that is a division of an **era**.

Phanerozoic Eon The geologic time span between 543 million years ago and the present; it embraces the **Paleozoic**, **Mesozoic**, and **Cenozoic** eras.

phaneritic A **textural** term used to describe coarse-grained **igneous** rocks. The grains are visible with the unaided eye.

phenocrysts Large crystals surrounded by smaller-grained matrix in an igneous rock.

phyllite A **foliated metamorphic** rock formed from **shale** with a **texture** between **slate** and **schist**. The rock generally consists of very fine-grained muscovite or chlorite whose **cleavage** give the rock a sheen when viewed in reflected light.

plastic deformation Rock deformation in which the mineral grains behave like plastic and, when compressed or sheared, become flattened or elongate without breaking.

plate tectonics A theory that horizontal movement of slabs of Earth's outer crust is responsible for the features of the Earth's surface.

plateau A relatively flat upland surface.

Pleistocene ice age A span of time in the geologic past from about 2 million years ago to about 9000 years ago when the Earth experienced an ice age.

plunging fold An **anticline** or **syncline** that is tilted so that it dips into the ground.

pluton An **igneous** intrusion.

Precambrian The span of geologic time between the Earth's origin about 4.6 billion years ago and the beginning of the **Phanerozoic Eon** about 542 million years ago.

pressure Force per unit area or the "push" or "pull" applied to material.

principle of cross-cutting relations When one geologic feature cuts across another; the feature that has been cut is the older.

principle of fossil succession In a sequence of **sedimentary rock**s, different species of **fossils** appear in a definite order; once a fossil species disappears in a sequence of strata, it never reappears higher in the sequence.

principle of inclusions When one rock contains fragments of another rock, the fragments and the rock from which they were derived must be older than the rock in which the fragments occur.

principle of continuity Sedimentary layers, before erosion, forming fairly continuous sheets over a region.

principle of original horizontality Holds that layers of sediment when originally deposited are relatively flat-lying.

principle or rule of sequence Placing geologic events in a chronological order based on elements of the rock record.

principle of superposition In a sequence of undisturbed sedimentary rocks, the lowermost layer is the oldest and the succeeding layers become progressively younger toward the top.

principle of uniformitarianism The geologic processes observed today also operated in the past in the same way and at comparable rates. According to James Hutton, "The present is the key to the past."

Proterozoic The most recent of the Precambrian eons.

protolith The original rock prior to metamorphism.

quarry An excavation from which rock or mineral resources are extracted.

quartz A mineral composed of pure silicon and oxygen (chemical formula SiO_2).

quartzite A nonfoliated metamorphic rock composed of quartz and whose protolith was a quartz sandstone. When fractured, it breaks through the cement and the grains, rather than around the grains as it would have in the protolith.

radioactive decay The process by which a radioactive isotope undergoes fission or release of particles.

radioactive isotope An unstable isotope of an element known to decay over a given time.

radiometric dating A method of absolute time dating geologic events in years by measuring the ratio of parent atoms to daughter atoms in materials containing radioactive isotopes.

regional metamorphism Dynamothermal metamorphism; metamorphism of a broad region, usually the result of deep burial during an orogeny.

regolith The loose unconsolidated material at the land surface. The usual parent material for soil.

regression The withdrawal of an epicontinental sea caused by the lowering of the sea level or the uplifting of the land.

relative age One method of dating geologic events. An event is either older or younger than a given datum.

relief The difference in elevation between adjacent high and low regions on the Earth's surface.

renewable resource A natural resource that can be replaced within a short span of time relative to a human life span.

reservoir rock A rock with sufficient porosity and permeability to contain concentrations of hydrocarbons.

reverse fault A steeply inclined fault in which the hanging wall apparently moved upward relative to the foot wall.

rhyolite A fine-grained, generally light-colored igneous rock with about 70 percent silica. (The extrusive equivalent of granite.)

rift valley A depression between mountain ranges in a continental rift zone.

rift zone A region experiencing pull-apart or divergence as within a plate or between plates.

rock A naturally occurring aggregate of one or more minerals, or a partially or wholly glassy rock that would have formed a mineral aggregate had it formed under different conditions.

rock cycle A series of events that results in the transformation of earth materials from one rock type to another. A continuous recycling of earth materials.

Rodinia A supercontinent that existed about 1 billion years ago resulting from the Grenville orogeny.

salt Many halide minerals belong in this group, but the most common is the mineral halite (table salt or sodium chloride). The crystal form and the cleavage of halite form perfect cubes.

salt hoppers The casts of former cubes of salt formed of sediment that filled the molds created by the dissolution of the salt.

sandstone A mechanically deposited sedimentary rock composed of sand-sized particles that are typically quartz but may be any composition.

saprolite The chemically weathered aspect of a rock that, although rotten in appearance, has retained some of its structural attributes.

schist A metamorphic rock resulting from relatively high temperature and pressure. Minerals in schist are large enough to identify. Schist is generally higher grade than phyllite but lower than gneiss.

seafloor spreading The gradual widening of an ocean basin as new crust emanates along a mid-ocean ridge (rift zone) and moves away from the ridge axis.

secondary enrichment A process whereby metals are leached from an original mineral deposit and redeposited below to form a higher grade ore deposit.

sedimentary basin A depression that fills with water and subsequently receives rock debris. eroded from adjacent uplands.

sedimentary rock A rock produced from material derived from preexisting rocks. The rock may be clastic, whereby the sediment is mechanically deposited, or nonclastic, whereby the sediment is deposited chemically, biologically, or biochemically.

sedimentary or clastic wedge An accumulation of sediment deposited on a subsiding passive continental margin.

sequence stratigraphy A means of dividing the stratigraphic record into unconformity-bound units of sedimentary rocks (sediment). Each unit begins with a time-transgressing deposit (typically a sandstone formed at the shore of the advancing sea) and ends with another time-transgressing unit (often a sandstone) that marks the regression of a sea. Furthermore these events are usually related to tectonic events.

shale A sedimentary rock composed of fine mud.

shatter cones Small cone-shaped fractures formed by shock induced by meteorite impact.

shield A region of the interior of a continental craton where old igneous and metamorphic rocks are exposed.

silica An ion consisting of the elements silicon and oxygen that combines with other elements to form the silicate minerals that comprise most of the Earth's crust and mantle.

silicate A group of minerals having the elements silicon and oxygen as the major constituents.

siliceous Rocks that are preponderantly composed of the mineral quartz.

sill A sheetlike concordant pluton.

siltstone A sedimentary rock composed of particles larger than clay-sized and smaller than sand-sized.

sinkhole A depression in the land surface in karst areas caused by the dissolution of the bedrock below it. Typically associated with limestone.

slate A foliated metamorphic rock formed from shale that was subjected to relatively small increases in temperature/pressure. (Slate is a lower grade metamorphism than phyllite.)

slaty cleavage metamorphic foliation The property of slate to break along planes that are not parallel to the bedding orientation of the original rock given the preferred orientation of clay minerals caused by deformation.

sorting 1. The range of clast sizes in a collection of sediment; 2. The degree to which sediment has been separated by flowing currents into different grain-size fractions.

stratigraphic column A diagram that illustrates the sequence of rock layers developed in a particular region of the Earth.

stratigraphy The branch of geology that deals with the establishment of a chronology of geologic events, especially with layered sedimentary rocks.

streak The color of a mineral's powder when pulverized.

stream A body of water flowing in a channel. Part of a drainage system.

stream capture (piracy) A condition whereby a headward eroding stream intercepts the water of another stream and, by virtue of having a more favorable gradient, the water is diverted into the headward eroding stream.

strike-slip fault A fault in which the blocks on either side of the break slid horizontally past one another when subjected to stresses.

stromatolite A layered mound of sediment formed by cyanobacteria; the cyanobacteria secrete a sticky substance that binds sediment into layers that become buried, and then this is followed by more colonizing bacteria and sediment that continue to build upward into a mound.

subaerial Above sea level or land surface.

subduction Where one plate slides beneath another at a convergent plate boundary.

suspect terrane See accreted terrane.

suture zone A region of deformation between two accreted terranes, between an accreted terrane and a continent, or between two continents that have collided.

syncline A fold in which rock layers are bent downward.

Taconic orogeny A convergent mountain-building event that occurred when a volcanic island arc collided with eastern proto-North America about 400 million years ago.

tectonics A study of the large features on Earth's surface and the internal processes that generated them.

tension A stress that pulls on material that can cause extension.

terrace An elevated surface of an old floodplain onto which a younger floodplain has been constructed by erosion and deposition. (See peneplain).

terrane A large region bounded by faults and that has a distinctive geology that is considerably different from the geology of the surrounding regions.

thin-skinned deformation A style of deformation characterized by displacement on thrust faults terminated along a subhorizontal detachment fault (called a décollement or sole fault).

thrust fault A low angle reverse fault where the hanging wall has apparently moved upward relative to the foot wall.

tidal flat A broad, nearly horizontal plain of mud and silt, exposed or nearly exposed at low tide, but totally submerged during high tide.

tillite A rock lithified from ancient glacial deposits consisting of large clasts embedded in a matrix of finer sand and mud.

topographic map A map drawn to scale that uses contours (lines of equal elevation) to represent the configuration of the land surface.

topography The configuration of the land surface.

transform fault The actively slipping segment of a fracture zone between two segments of a spreading center (e.g., mid-ocean ridge).

transform plate boundary A region where two masses of the Earth's outer shell (lithosphere) slide laterally past one another.

transgression The advance of a sea overland as a result of land subsidence or rising sea level.

trellis network or drainage pattern A drainage pattern marked by streams flowing in parallel valleys with tributaries heading on adjacent ridges and intersecting the main streams at nearly right angles. Resembles a garden trellis.

trench A deep elongate trough bordering a volcanic arc; a trench defines the trace of a convergent plate boundary.

turbidite A graded bed of sediment built up at the base of a submarine slope and deposited by turbidity currents.

turbidity current A submarine avalanche of sediment and water that, as a result of its density, flows with great velocity while not mixing with the less dense surrounding water.

ultimate base level Sea level. The level below which a trunk stream cannot erode downward.

unconformity A surface in the rock record for which time cannot be accounted. Either no sediment was deposited or sediment was deposited and subsequently eroded away. See disconformity, nonconformity, and angular unconformity.

varve A pair of thin layers of sediment formed in a glacial-lake bed. One layer is coarser that the other and is interpreted to have been deposited during summer when the lake was agitated by wind and stream flow; the other is fine and believed to have been deposited during the winter when ice covered the lake.

volcanic arc A curving chain of active volcanoes formed adjacent to a converging plate boundary.

volcanic island arc A volcanic island formed when one oceanic plate subducts beneath another ocean plate.

waterfall A place where a river or stream drops over a ledge of resistant rock.

water gap A cut through a resistant ridge where a river or stream has eroded its channel.

watershed A region that collects the atmospheric precipitation that feeds a given drainage network.

weathering A process at the Earth's surface whereby Earth materials are chemically decomposed into different products or whereby the materials are physically disintegrated into smaller and smaller fragments. Physical weathering commonly facilitates chemical weathering.

Bibliography

Adams GI. 1928 The course of the Tennessee River and the physiography of the Southern Appalachian region. Jour. Geology. 36: 481–493.

Ashley GH. 1911. The gold fields of Coker Creek, Monroe County, Tennessee. Tenn. Geol. Survey, Res. Tenn. Vol. 1, p. 78–107.

Barr TC. 1961. Caves of Tennessee. Tenn. Div. Geology Bull. 64. 567 p.

Bartholomew MJ, Lewis SE. 1984. Evolution of Grenville massifs in the Blue Ridge geologic province, Southern and Central Appalachians. In Bartholomew MJ and others, eds., The Grenville event in the Appalachians and related topics. Boulder, CO: Geological Society of America Spec. Paper 194, p. 229–254.

Bayley WS. 1923. The magnetic iron ores of East Tennessee and Western North Carolina. Tenn. Div. Geology Bull. 29. 252 p.

Bird JM, Dewey JF. 1970. Lithosphere plate-continental margin tectonics and the evolution of the Appalachian orogen. Geol. Society of America Bull. 81: 1031–1061.

Blythe Jr EW. 1986. Field trip in the Reelfoot Lake area, Tennessee. In Neathery TL, ed. Geological Society of America Decade of North American Geology, Centennial Field Guides, southeastern section.

———, McCutchen WT, Stearns RG. 1975. Geology of Reelfoot Lake and vicinity. Tenn. Div. of Geology Rept. Inv. No. 36, p. 64–76.

Byerly DW. 1966. Structural geology along a segment of the Pulaski fault, Greene County, Tennessee. Doctoral dissertation, Dept. of Geological Sciences, University of Tennessee. 94 p.

Byerly DW, Walker KR, Diehl WW, Ghazizadeh M, Johnson RE, Lutz CT, Schoner AK, Simmons WA, Simonson JCB, Weber LJ, Wedekind JE. 1986. Thorn Hill: a classic Paleozoic stratigraphic section in Tennessee. Geol. Society of America Centennial Field Guide. Southeastern sect., p. 131–136.

Carpenter RH. 1970. Metamorphic history of the Blue Ridge province of Tennessee and North Carolina. Geol. Society of America Bull. 81: 749–761.

Caudill, M., Driese SG, Mora CI. 1996. Preservation of a paleo-Vertisol and an estimate of Late Mississippian paleoprecipitation. Jour. Sedimentary Research. 66: 58–70.

Chew C. 1997. A geologic history of Bays Mountain Park. 3rd. ed. Kingsport, TN: NA Wink. 71 p.

Clark GM. 1989. Central and Southern Appalachian water and wind gap origins: review and new data. Geomorphology 2: 209–232.

Clark SHB. 2009. Birth of the mountains: the geologic story of the Southern Appalachian Mountains. U.S. Geological Information Services. 23 p.

Conant LC, Swanson VE. 1961. Chattanooga Shale and related rocks of Central Tennessee and nearby areas. U.S. Geol. Survey Prof. Paper 357. 91 p.

Corgan JX. 1985. Early medical dissertation research in Tennessee geology. Geol., Earth Sciences and History of Earth Science. 4(1): 10–16.

Corgan JX, Parks JT. 1979. Natural bridges of Tennessee. Tenn. Div. Geology Bull. 80. 102 p.

Corgan JX, Gibson MA. 1991. Early scientific exploration of West Tennessee. Gerard Troost's travels in 1834. West Tennessee Historical Society Journal. 45: 83–94.

———. 1996 Geological exploration in East Tennessee. Gerard Troost's travels in 1834, Journal of the East Tennessee Historical Society. 54(2): 140–153.

Cudzil MR, Driese SG. 1987. Fluvial, tidal, and storm sedimentation in the Chilhowee Group (Lower Cambrian), Northeastern Tennessee. Sedimentology. 34: 861–883.

Deane, B., Lee, P., Milam KA, Evenick JC, Zawislak RL. 2004. The Howell Structure, Lincoln County, Tennessee: a review of past and current research. Lunar and Planetary Science. 35 [cited 14 Sept. 2012]. 2 p. Available online: http://www.lpi.usra.edu/meetings/lpsc2004/pdf/1692.pdf.

Delcourt PA. 1980. Quaternary alluvial terraces of the Little Tennessee River Valley, East Tennessee. In Chapman, J., ed., The 1979 archeological and geological investigation in the Tellico Reservoir. University of Tennessee Dept. of Anthropology Rept. 29, p. 110–121.

Diamond J. 2005 Collapse: how societies choose to fall or succeed. New York: Penguin Group. 575 p.

Dietz RS. 1959. Shatter cones in cryptoexplosion structures (meteorite impact?). Jour. Geology. 67: 496–505.

Driese SG. 1988. Depositional history and facies architecture of a Silurian foreland basin, Eastern Tennessee. In Driese SG, Walker D, eds., Depositional history of Paleozoic Sequences, Southern Appalachians. University of Tennessee, Department of Geological Sciences Studies in Geology 19, p. 62–94.

Dunbar CO. 1919. Stratigraphy and correlation of the Devonian of Western Tennessee. Tenn. Div. Geology Bull 21. 127 p.

Dunigan T. 2011. Tennessee waterfalls [cited 2012 Sept 3]. Available from http://web.eecs.utk.edu/~dunigan/landforms/falls.php.

Emmons WH, Laney FB. 1926. Geology and ore deposits of the Ducktown mining district, Tennessee. U.S. Geol. Survey Prof. Paper 139. 114 p.

Evenick JC, Hatcher Jr RD. 2006. Geophysically subdividing the Nashville (Trento) and Stones River (Black River) Groups beneath the Eastern Highland Rim and Southern Cumberland Plateau in Tennessee and Southern Kentucky. Tenn. Div. Geology Rept. Inv. No. 52. 15 p.

Fenneman NM. 1928. Physiographic divisions of the United States, Annals of the Assoc. of American Geographers,, 18, p. 261–353.

———. 1931. Physiography of the western United States. New York, McGraw-Hill. 534 p.

———. 1938. Physiography of the eastern United States. New York, McGraw-Hill. 714 p.

Ferm JC. 1974. Carboniferous environmental models in eastern United States and their significance. Geol. Society of America Spec. Paper 148. 16 p.

Ferm JC, Horne JC, Swinchatt JP, Whaley PW. 1971. Carboniferous depositional environments in northeastern Kentucky. In Kentucky Geol. Society Guidebook. Ann. Spring Field Conf., Lexington, KY. 30 p.

Ferm JC, Milici RC, Eason JE. 1972. Carboniferous depositional environments in the Cumberland Plateau of Southern Tennessee and Northern Alabama. Tenn. Div. Geology Rept. Inv. 33. 32 p.

Fisher GW, Pettijohn FJ, Reed Jr JC, and Weaver KN. 1975. Studies in Appalachian geology: central and southern. New York: John Wiley–Interscience. 460 p.

Floyd RJ. 1965. Tennessee rock and mineral resources. Tenn. Div. Geology Bull. 66. 119 p.

Fox PP, Grant LF. 1944. Ordovician bentonites in Tennessee and adjacent states. Jour. Geology. 52: 319–332.

Fullagar PD, Bartholomew MJ. 1983. Rubidium-strontium ages of the Watauga River, Cranberry, and Crossing Knob gneisses, northwestern North Carolina. In Lewis SE, ed., Geological investigations in the Blue Ridge of northwestern North Carolina. Guidebook for the Carolina Geological Society, North Carolina Division of Land Resources, Article II. 29 p.

Fuller ML. 1912. The New Madrid earthquake. U.S. Geol. Survey Bull. 494. 119 p.

Gibson MA. 1988. Paleogeography, depositional environments, paleoecology, and biotic interactions of the Rockhouse Limestone and Birdsong Shale Members of the Ross Formation (Lower Devonian), Western Tennessee. Doctoral dissertation, Dept. of Geol. Science, University of Tennessee.

——. 1990. Paleogeography and paleobiography of the Lower Devonian, southeastern U.S. Geol. Soc. of America, Southeastern Section Meeting Abstracts with Program. 22(4): 16.

Glenn LC. 1915. Physiographic influence in the development of Tennessee. Tenn. Div. Geology. Resources of Tennessee. 5(2): 44–64.

Goldberg SA, Butler JR, Fullagar PD. 1986. The Bakersville dike swarm: geochronology and petrogenesis of Late Proterozoic basaltic magmatism in the Southern Appalachian Blue Ridge. Amer. Jour. Science. 286(5): 403–430.

Goldberg SA, Dallmeyer RD. 1997. Chronology of Paleozoic metamorphism and deformation in the Blue Ridge thrust complex, North Carolina and Tennessee. Amer. Jour. Science. 297(5): 488–526.

Gordon CH. 1924. Marble deposits of East Tennessee: occurrence and distribution. Tenn. Div. Geology Bull. 28: 15–86.

Hack JT. 1966. Interpretation of Cumberland Plateau escarpment and Highland Rim, South-Central Tennessee and Northern Alabama. U.S. Geol. Survey Prof. Paper 524 C, p. C1–C16.

Hadley JB, Goldsmith R. 1963 Geology of the Eastern Great Smoky Mountains, North Carolina and Tennessee. U.S. Geol. Survey Prof. Paper 349-B. 118 p.

Hale, Robin C., 1990, The gold fields of Coker Creek, Monroe county, Tennessee. Tennessee Div. Geol. Bull. 72. 93 p.

Hall GM, Amick HC. 1944. Igneous rock areas in the Norris region. Tennessee. Jour. of Geology. 52(6): 424–430.

Hamilton W. 1963 Geology of the Richardson Cove and Jones Cove Quadrangle, Tennessee. U.S. Geol. Survey Prof. Paper 349-A. 55 p.

Hardeman WD. 1966. Geologic map of Tennessee, scale 1:250,000. Nashville: Tenn. Div. Geology.

Harris LD. 1964. Facies relations of exposed Rome Formation and Conasauga Group of Northeastern Tennessee with equivalent rocks in the subsurface of Kentucky and Virginia. U.S. Geol. Survey Prof. Paper 501-B, B25–B29.

——. 1969. Kingsport Formation and Mascot Dolomite (Lower Ordovician) of East Tennessee. Tenn. Div. Geology Rept. Inv. 23. 39 p.

Harris LD, Milici RC. 1977. Characteristics of thin-skinned style of deformation in the southern Appalachians and potential hydrocarbon traps. U.S. Geol. Survey Prof. Paper 1018. 40 p.

Hass WH. 1956. Age and correlation of the Chattanooga Shale and the Maury Formation. U.S. Geol. Survey Prf. Paper 286. 47 p.

Hasson KO, Haase CS. 1988. Lithofacies and paleogeography of the Conasauga Group (Middle and Late Cambrian) in the Valley and Ridge Province of East Tennessee. Geol. Society of America Bull. Vol. 100, p 234–246.

Hatcher Jr RD, Merschat CE, Milici RC, Wiener LS. 1978. A structural transect in the southern Appalachians, Tennessee and North Carolina. In Milici RC, chairman, Field trips in the Southern Appalachians. Tenn. Div. Geology Rept. Inv. No. 37, p 6–51.

Hatcher Jr RD, Thomas WA, Viele GW, eds. 1989. The Appalachian-Ouachita Orogen in the United States. The Geology of North America. Vol. F-2. Boulder, CO: Geological Society of America. 767 p.

Hershey RE, Maher SW. 1985. Limestone and dolomite resources of Tennessee. 2nd Ed., Tenn. Div. Geology Bull. 65. 251 p.

Horton JW, Zullo VA, eds., 1991, The geology of the Carolinas. Knoxville: University of Tennessee Press. 406 p.

Hurst VJ, Schlee JS. 1962. Field excursion. Ocoee metasediments, north-central Georgia, southeastern Tennessee. Georgia Geological Survey Guidebook 3. 28 p.

Hutton, J. 1785. Systems of the earth, its duration, and stability, Abstract of a dissertation delivered to the Royal Society of Edinburgh, Scotland.

———. 1795. Illustrations of the earth, with proofs and illustrations, 2 vols. Edinburgh, Scotland: Creech Book Publisher.

Johnson DW. 1941. Mussel distribution as evidence of drainage changes. Jour. Geomorphology. 4: 307–321.

———. 1942. Mussel distribution as evidence of drainage changes. Jour. Geomorphology. 5: 59–72.

Keith A. 1895. Description of the Knoxville sheet, North Carolina–Tennessee. U.S. Geological Survey. Geol. Atlas, Cranberry Folio 16. 6 p.

———. 1903. Description of the Cranberry Quadrangle, North Carolina–Tennessee. U.S. Geological Survey. Geol. Atlas, Cranberry Folio 90. 9 p.

———. 1904. Description of the Asheville Quadrangle, North Carolina–Tennessee. U.S. Geological Survey. Geol. Atlas, Cranberry Folio 116. 10 p.

———. 1907. Description of the Roan Mountain Quadrangle, Tennessee–North Carolina. U.S. Geological Survey. Geol. Atlas, Cranberry Folio 116. 10 p.

Kellberg JM, Grant LF. 1956. Coarse conglomerates of the Middle Ordovician in the Southern Appalachian Valley. Geol. Society of America Bull. 67: 697–716.

Keller FB. 1980. Later Precambrian stratigraphy, depositional history and structural chronology of part of the Tennessee Blue Ridge. Ph.D. thesis. Yale University. 353 p.

King PB. 1964. Geology of the Central Great Smoky Mountains, Tennessee. U.S. Geol. Survey Prof. Paper 349-C. 148 p.

King PB, Hadley JB, Neuman RB, Hamilton W. 1958. Stratigraphy of Ocoee Series, Great Smoky Mountains, Tennessee and North Carolina. Geol. Society of America Bull. 69, p 947–966.

King PB, Ferguson HW. 1960. Geology of northeasternmost Tennessee. U.S. Geol. Survey Prof. Paper 311. 136 p.

King PB, Neuman RB, Hadley JB. 1968. Geology of the Great Smoky Mountains National Park, Tennessee and North Carolina. U.S. Geol. Survey Prof. Paper 587. 23 p.

Kish SA, ed. 1991, Studies of Precambrian and Paleozoic stratigraphy in the western Blue Ridge. Carolina Geological Society Field Trip, November 9–10, 1991. 161 p.

Kohl, M. 2002. Geology of Norris Dam State Park. Tenn. Div. Geology, State Park Series 2. 17 p.

Lemiszki P J, 2003, Geology of Roan Mountain State Park. Tenn. Div. Geology, State Park Series 3. 21 p.

Lillie RJ. 2005. Parks and plates: the geology of our national parks. New York/London: W.W. Norton and Company. 298 p.

Luppens JA. 1980. Gulf Coast lignite trends: distribution, quality, and reserves. Houston: Gulf Coast Lignite Conf. Proc.

Luther, ET. 1977. Our restless earth: the geologic regions of Tennessee. Knoxville: University of Tennessee Press. 94 p.

Marcher MV, Stearns RG. 1962. Tuscaloosa formation in Tennessee. Tenn. Div. Geology Rept. Inv. 17. 22 p.

McAdoo WG, White, 1861, Elementary geology of Tennessee: for use of the public schools and other institutions of learning. Cincinnati, OH: American Book Co. 118 p.

McPhee J. 1998. Annals of the former world. New York: Farrar, Straus, and Giroux. 696 p.

Merschat CE, Wiener LS. 1990. Geology of Grenville-Age basement and younger cover rocks in the west central Blue Ridge, North Carolina. Carolina Geol. Society Field Trip Guidebook, North Carolina Geological Survey, Raleigh NC. 42 p.

Meyer HOA. 1975. Kimberlite from Norris Lake, eastern Tennessee. Mineralogy and Petrology. Jour. of Geology. 83: 518–526.

Miller RA. 1974. The geologic history of Tennessee. Tenn. Div. Geology Bull. 74. 63 p.

Milici RC. 1963. Low-angle overthrust faulting as illustrated by the Cumberland Plateau–Sequatchie Valley fault system. Amer. Jour. Science. 261: 815–825.

———. 1968a. The physiography of Sequatchie Valley and adjacent portions of the Cumberland Plateau, Tennessee. Tenn. Div. Geology Rept. Inv. 22. 15 p.

———. 1968b. Mesozoic and Cenozoic physiographic development of the lower Tennessee River: in terms of the dynamic equilibrium concept. Jour. Geology. 76: 472–479.

———. 1970a. The Allegheny structural front in Tennessee and its regional implications. Amer. Jour. Science. 268: 127–141.

———. 1970b. Middle Ordovician stratigraphy in Central Sequatchie Valley, Tennessee. Tenn. Div. Geology Rept. Inv. 30. 17 p.

———. 1973. Geology of Knox County Tennessee: the stratigraphy of Knox County, Tennessee. Tenn. Div. Geology Bull. 70. 9–24.

———. 1974. Stratigraphy and depositional environments of Upper Mississippian and Lower Pennsylvanian rocks in the Southern Cumberland Plateau of Tennessee. In Briggs G, ed., Carboniferous of the Southeastern United States. Geological Society of America Spec. Paper 148, p. 115–133.

Milici RC, Smith JW. 1969. Stratigraphy of the Chickamauga Supergroup in its type area. Tenn. Div. Geology Rept. Inv. 24. 35 p.

Milici RC, de Witt W. 1988. The Appalachian Basin. In: Sloss LL, ed. Sedimentary cover— North American Craton: U.S. Boulder, CO: Geological Society of America. p. 427–469.

Mills HH, Kaye JM. 2001. Drainage history of the Tennessee River: review and new metamorphic quartz gravel locations. Southeastern Geology. 40(2): 75–97.

Monrad JR, Gulley GL. 1983. Age and P-T conditions during metamorphism of granulite— facies gneisses, Roan Mountain, NC-TN. In: Lewis SE, ed., Geologic investigations in the Blue Ridge of northwestern North Carolina. Carolina Geological Society Field Trip Guidebook. 29 p.

Moore HL. 1992. A roadside guide to the geology of the Great Smoky Mountains National Park. Knoxville: University of Tennessee Press. 178 p.

———. 1994. A geologic trip across Tennessee by Interstate 40. Knoxville: University of Tennessee Press. 339 p.

Mulholland JW. 1998. Sequence stratigraphy: basic elements, concepts, and terminology. Australian Soc. Exploration Geophysics. 73: 1–7.

Neuman RB. 1955. Middle Ordovician rocks of the Tellico-Sevier Belt, eastern Tennessee. U.S. Geol. Survey Prof. Paper 274-F, p. 114–178.

Neuman RB, Nelson WH. 1963. Geology of the Western Great Smoky Mountains, Tennessee. U.S. Geol. Survey Bull 349-D. 81 p.

Odom AL, Fullagar PD. 1984. Rb-Sr whole-rock and inherited zircon ages of the plutonic suite of the Crossnore Complex, Southern Appalachians, and their implications regarding the time of opening of the Iapetus Ocean. Geol. Society of America Spec. Paper 194, p. 255–261.

Plum GA. 2009. Waterfalls of Tennessee. Johnson City, TN: Overmountain Press. 287 p.

Potter PE. 1955. The petrology and origin of the Lafayette gravel. Part 2: Geomorphic history. Jour. Geology. 63: 115–132.

Rankin DW. 1993. The volcanogenic Mount Rogers Formation and the overlying Konnarock Formation—two Late Proterozoic units in Southwestern Virginia. U.S. Geol Survey Bull. 2029. 26 p.

Rast N. 1989. The evolution of the Appalachian chain. In: Geology of North America. Vol. A, The geology of North America—an overview. Boulder, CO: Geol. Society of America. p. 323–348.

Rast N, Kohles KM. 1986. The origin of the Ocoee Supergroup. Amer. Jour. Science. 286: 593–616.

Reesman AL, Stearns RG. 1989. The Nashville Dome—an isostatically induced erosional structure and the Cumberland Plateau Dome—an isostatically suppressed extension of the Jessamine Dome. Southeastern Geology 30: 147–174.

Reichert SO. 1942. Manganese resources of East Tennessee, with partial reprinting of USGG Bull. 737, edited by Whitlach GI. Tenn. Div. Geol. Bull. 50. 212 p.

Rich JL, 1934. Mechanics of low-angle overthrust faulting as illustrated by the Cumberland thrust block, Virginia, Kentucky, and Tennessee. Amer. Assoc. Petroleum Geologists Bull. 18(12): 1634–1654.

Robinson Jr GR, Lesure JI, Marlowe II FNK, Clark SH. 1992. Bedrock geology and mineral resources of the Knoxville 1° x 2° Quadrangle, Tennessee, North Carolina, and South Carolina. U.S. Geol. Survey Bull. 1979. 73 p.

Roddy DJ. 1966. The Paleozoic crater at Flynn Creek, Tennessee. Doctoral dissertation. Calif. Inst. Tech.

———. 1968. Shock metamorphism in carbonate rocks at probable impact structures (abstract). Geological Society of America, Cordilleran Section, 64th Annual Meeting, Tucson, Arizona. P. 103.

Rodgers J. 1953 Geologic map of East Tennessee with explanatory text. Tenn. Div. Geol. Bull. 58, part 2. 168 p.

———. 1970. The tectonics of the Appalachians. New York: Wiley–Interscience Publishers. 271 p.

Rodgers, J., and Kent DF. 1948. Stratigraphic section at Lee Valley, Hawkins County, Tennessee. Tenn. Div. Geology Bull. 55. 47 p.

Ruppel SC, Walker KR. 1984. Petrology and depositional history of a Middle Ordovician carbonate platform. Chickamauga Group, northeastern Tennessee. Geol. Society of America Bull. 95: 568–583.

Russell EE, Parks WS. 1975. Stratigraphy of the outcropping Upper Cretaceous, Paleocene, and Lower Eocene of Western Tennessee (Including descriptions of younger fluvial deposits). Tenn. Div. Geology Bull 75. 128 p.

Safford JM. 1869. Geology of Tennessee. Nashville: SC Mercer, printer. 550 p.

Safford JM, Killebrew JB. 1885. The elementary geology of Tennessee: being also an introduction to geology in general designed for schools of Tennessee: Nashville: AB Tavel. 255 p.

Secrist MH. 1924. Zinc deposits of East Tennessee. Tenn. Div. Geology Bull. 31. 165 p.

Schultz AP, Southworth S. 2000. Geology, Great Smoky Mountains National Park. Great Smoky Mountains Natural History Association, geologic map and text.

Self RP. 2000. The pre-Pliocene course of the Lower Tennessee River as deduced from river terrace gravels in Southwest Tennessee. Southeastern Geology. 39: 61–70.

Servais T, Harper DAT, Jun Li MA, Owen AW, Sheehan PM. 2009. Understanding the great Ordovician biodiversification event (GOBE): influence of paleogeography, paleoclimate, or paleoecology? GSA Today 19(4/5): 4–10.

Shanmugam G, Walker KR. 1980. Sedimentation, subsidence, and evolution of a foredeep basin in the Middle Ordovician, Southern Appalachians. Amer. Jour. Science. 280: 479–496.

Shunk AJ, Driese SG, Clark GM. 2006. Latest Miocene to earliest Pliocene sedimentation and climate record derived from paleosinkhole fill deposits, Gray Fossil Site, Northeastern Tennessee. Paleo. 231(3-4): 265–278.

Sloss LL. 1963. Sequences in the cratonic interior of North America. Geological Society of America Bull. No. 2, p. 93–114.

Stearns RG. 1954. The Cumberland Plateau overthrust and geology of the Crab Orchard Mountains area, Tennessee. Tenn. Div. Geology Bull. 60. 47 p.

———. 1958. Cretaceous, Paleocene, and Lower Eocene geologic history of the northern Mississippi Embayment. Tenn. Div. Geology Rept. Inv. 6. 24 p.

———, ed. 1975. Field trips in West Tennessee. Tenn. Div. Geology Rept. Inv. 36. 82 p.

Stearns RG, Armstrong CA. 1955. Post-Paleozoic stratigraphy of Western Tennessee and adjacent portions of the Upper Mississippi Embayment. Tenn. Div. Geology Rept. Inv. 2. 29 p.

Stearns RG, Marcher MV. 1962. Later Cretaceous and subsequent structural development of the Northern Mississippi Embayment area. Tenn. Div. Geology Rept. Inv. 18. 8 p.

Stearns RG, Blythe Jr EW, Hoyal ML. 1989. Guidebook of geology and related history of the Chickasaw Bluffs and Mississippi River Valley in West Tennessee. Tenn. Div. Geology Rept. Inv. 45. 44 p.

Stose GW, Stose AJ. 1944. The Chilhowee Group and Ocoee Series of the Southern Appalachians. Amer. Jour. Science. 242: 367–390, 401–416.

Swingle GD. 1959. Geology, mineral resources and groundwater of the Cleveland area, Tennessee. Tenn. Div. Geology Bull. 61.

Su Q, Goldberg SA, Fullagar PD. 1994. Precise U-Pb zircon ages of Neoproterozoic plutons in the Southern Appalachian Blue Ridge and their implications for the initial rifting of Laurentia. Precambrian Research. 68: 81–95.

Tennessee Division of Geology. 1978. Field trips in the Southern Appalachians. Tenn. Div. Geology Rept. Inv. 37. 86 p.

Thomas WA. 1991. The Appalachian-Ouachita rifted margin of southeastern North America. Geol. Soc. Amer, Bull., v. 103, p. 415–431

Troost G. 1837. Fourth geological report to the twenty-second General Assembly of the State · of Tennessee, Nashville, October 1837. 37 p.

Upham GA. 1992. Tennessee mineral annual. Tennessee. Div. Geol. Bull. 83. 67 p.

——. 1993. Historical overview of Tennessee's mineral industry: past and present. Jour. Tenn. Acad. Science. 68(1): 23–31.

Van Arsdale R. 2009. Adventures through deep time: the central Mississippi River Valley and its earthquakes: the Geol. Society of America Special paper 455. 107 p.

Van Arsdale R, Cox R. 2007 Jan. The Mississippi's curious origins. Scientific American. P 33–39.

Vogt PR, Jung W-Y. 2007. Origin of the Bermuda volcanoes and Bermuda Rise: history, observations, models, and puzzles. In Foulger GR and Jurdy DM, eds., Plates, plumes, and planetary processes. Geological Society of America Special Paper 430, p 553–591.

Wade B. 1926 The fauna of the Ripley Formation on Conn Creek, Tennessee. U.S. Geol. Survey Prof. Paper 137. 272 p.

Walker JD, Geisman JW. 2009. Geologic time scale. Geol. Soc. America, doi: 10.1130/2009.

Walker KR, Ferrigno KF. 1973. Major Middle Ordovician reef tract in East Tennessee. Amer. Jour. Science. Cooper 273-A: 294–325.

Walker KR, Broadhead TW, Keller FB. 1980. Middle Ordovician carbonate shelf to deep water · basin deposition in the Southern Appalachians. University of Tennessee Studies in Geology no. 4, Dept. of Geological Sciences, University of Tennessee. 120 p.

Walker KR and others. 1985. The geologic history of the Thorn Hill Paleozoic (Cambrian-Mississippian), Eastern Tennessee. University of Tennessee Studies in Geology no. 10, Dept. of Geological Sciences, University of Tennessee. 128 p.

Wheeler HE. 1958. Time-stratigraphy. Amer. Assoc. Petroleum Geologist. 42: 1047–1063.

Whitlach GI. 1940. The clays of West Tennessee. Tenn. Div. Geology Bull. 49. 368 p.

Wilcox TE, Poldervaart A. 1958 Metadiorite dike swarm in Bakersville–Roan Mountain area, North Carolina. Geol. Society of America Bull. 69: 1323–1368.

Wilson CW Jr. 1948. The geology of Nashville, Tennessee. Tenn. Div. Geology Bull. 53. 172 p.

——. 1949. Pre-Chattanooga stratigraphy in Central Tennessee. Tenn. Div. Geology Bull. 56. 407 p.

——. 1953. Annotated bibliography of the geology of Tennessee through Dec. 1950. Tenn. Div. Geology Bull. 59. 308 p.

——. 1958. Guidebook to geology along Tennessee highways. Tenn. Div. Geology Rept. Inv. 5.

——. 1962. Stratigraphy and geologic history of Middle Ordovician rocks of Central Tennessee. Tenn. Div. Geology Rept. Inv. 15. 24 p.

——. 1973. Annotated bibliography of Tennessee geology, Jan. 1961 through Dec. 1970. Tenn. Div. Geology Bull. 71.

——. 1981. State geological surveys and state geologists of Tennessee. Tenn. Div. Geology Bull. 81. 62 p.

Wilson CW Jr, Jewell JW, Luther ET. 1956. Pennsylvanian geology of the Cumberland Plateau. Tenn. Div. Geology, Folio. 221 p.

Wilson CW Jr, Stearns RG. 1958. Structure of the Cumberland Plateau, Tennessee. Geol. Society of America Bull. 69(10): 1283–1296.

——. 1968. Geology of the Wells Creek structure, Tennessee. Tenn. Div. Geology Bull. 68.

Zurawski RP. 1981. Tennessee production trends on the rise. Northeast Oil Reporter. 1(4): 74–78.

Index